The Solar System

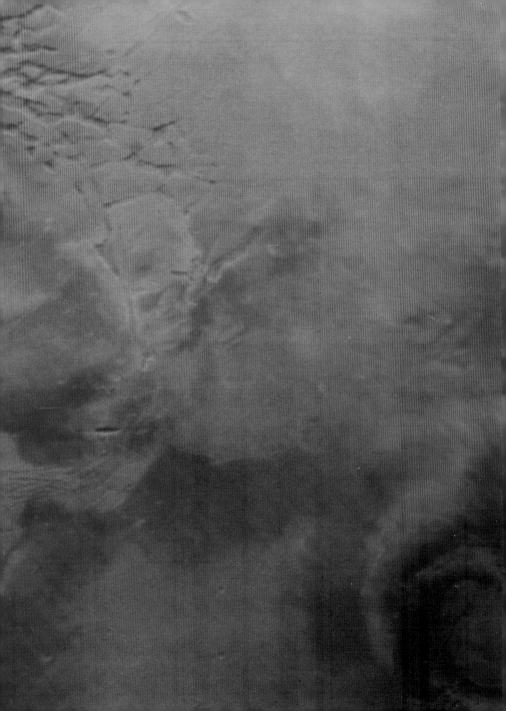

The Solar System

edited by **Giovanni Caprara**

FIREFLY BOOKS

Illustrations: p.1: the Lunar
Excursion Module prepares to land
on the Moon
pp. 2/3: surface of Mars
pp. 4/5: Jupiter
pp. 8/9: surface of Mercury

Text by:
Alessandro Braccesi, Margherita
Hack, Giovanni Caprara

Art Director
Giorgio Seppi

Coordination and editorial
production:
Studio Pleiadi, Cesena

Published in Canada in 2003 by
Firefly Books Ltd.
3680 Victoria Park Avenue
Toronto, Ontario M2H 3K1

Published in the United States in
2003 by
Firefly Books (U.S.) Inc.
P.O. Box 1338
Ellicott Station
Buffalo, New York 14205

Printed in Spain

D.L. TO: 825 - 2003

A Firefly Book

Published by Firefly Books Ltd. 2003

Copyright © 2001 Arnoldo Mondadori
Editore S.p.A.
English translation © Arnoldo
Mondadori Editore S.p.A.

Translated by S.M. Harris
Science content advisor for English
language edition: Tom Watters,
National Air and Space Museum,
Smithsonian Institution

All rights reserved. No part of this
publication may be reproduced,
stored in a retrieval system in any
form nor by any means, electronic,
mechanical, photocopying, recording
or otherwise, without the prior writ-
ten permission from the Publisher.

First printing 2002.

**Publisher Cataloguing-in-
Publication Data (U.S.)**
(Library of Congress Standards)

Solar system : a Firefly guide /
edited by Giovanni Caprara.—1st ed.
[256] p. : col. ill., photos. ; cm.
Includes bibliographical references
and index.

Originally published as "Sistema
Solare", Italy: Mondadori, 2001.
Summary: A comprehensive guide to
the solar system including planets,
stars, comets, asteroids, meteorites
and the sun.

ISBN 1-55297-679-3 (pbk.)
1. Solar system. I. Caprara,
Givovanni. II. Title
523.2 21   QB501.2.C37 2003

**National Library of Canada
Cataloguing-in-Publication Data**

The solar system : a Firefly guide /
edited by Giovanni Caprara ;
translated by S.M. Harris

Translation of: Sistema solare.
Includes bibliographical references
and index.

ISBN 1-55297-679-3
1. Solar system—Popular works.
I. Caprara, Giovanni II. Harris,
S.M. III. Title.
QB501.2.S5813 2003   523.2
C2002-904102-3

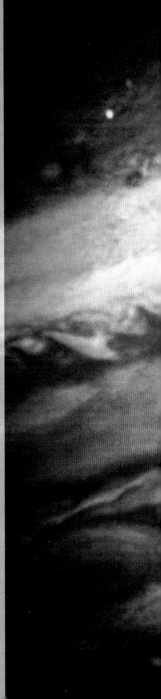

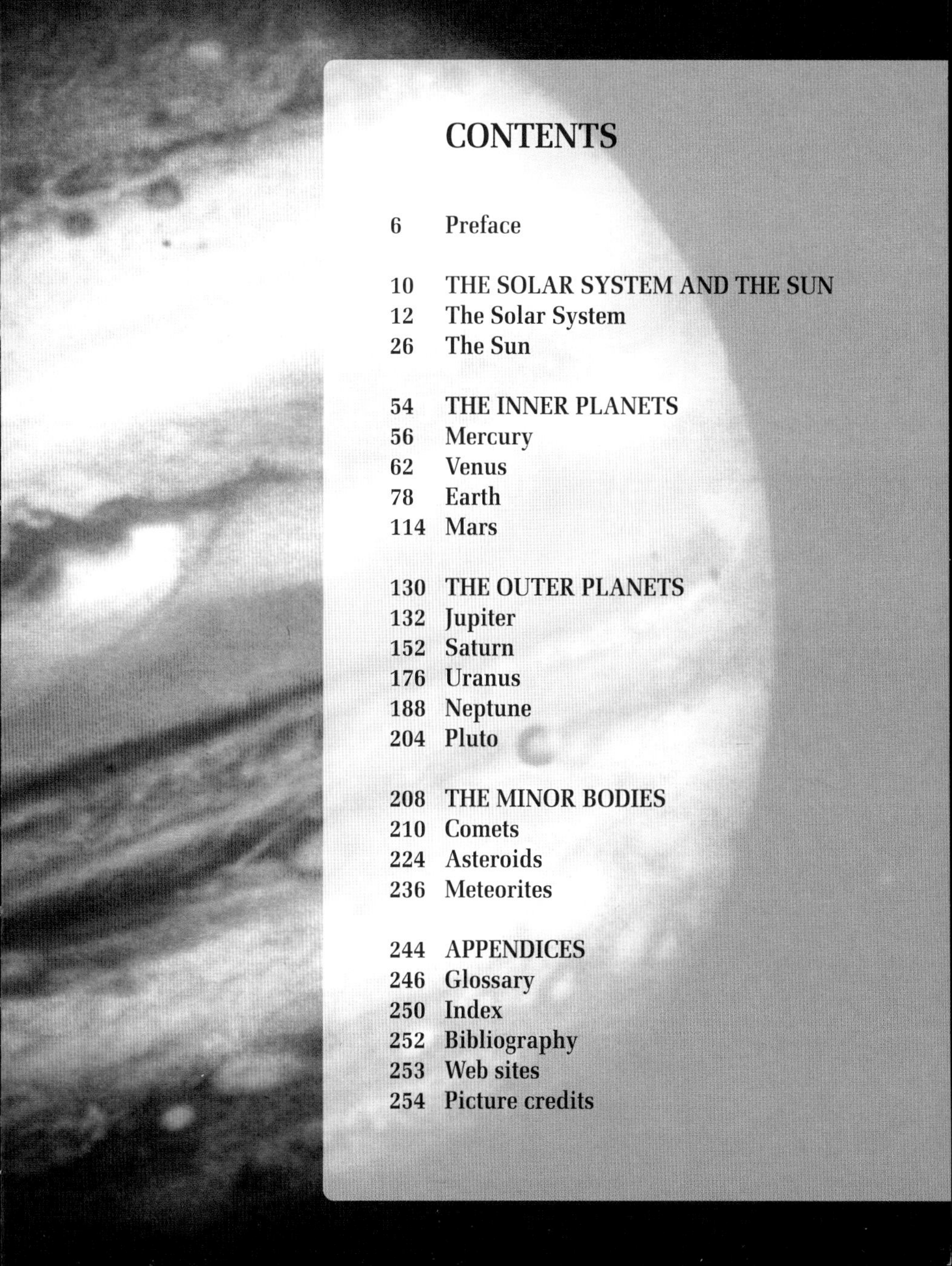

CONTENTS

6 Preface

10 **THE SOLAR SYSTEM AND THE SUN**
12 The Solar System
26 The Sun

54 **THE INNER PLANETS**
56 Mercury
62 Venus
78 Earth
114 Mars

130 **THE OUTER PLANETS**
132 Jupiter
152 Saturn
176 Uranus
188 Neptune
204 Pluto

208 **THE MINOR BODIES**
210 Comets
224 Asteroids
236 Meteorites

244 **APPENDICES**
246 Glossary
250 Index
252 Bibliography
253 Web sites
254 Picture credits

PREFACE

No one could fail to be stirred by memories of the panoramic views of Mars transmitted by the Mars *Pathfinder* probe which landed on the Red Planet with the Sojourner mini-robot in July 1997. The snow-white, icy surface of Europa, Jupiter's mysterious moon proved to be just as spell-binding when seen through the lens of the *Galileo* probe, as were the sinuous lines of the Eros asteroid which NASA's *NEAR* probe was finally able to observe in the minutest detail. The sight of the Mir space station, transformed into a 140-ton falling star, as it plunged into the Pacific ocean after so many years of invaluable service was equally unforgettable. These are the latest protagonists in the "second epoch" of planetary science, which began with the arrival of the space age at the end of the 1950s. Until then we were still in the "First epoch," which had begun with the great discoveries of the sixteenth and seventeenth centuries and in which the key players were four great figures in astronomy: Copernicus, Galileo, Kepler and Newton.

Changes in the technology at our disposal has revolutionized and hugely expanded our knowledge of the planets around our own, domestic star, and we have also learned much more about the Earth on which we are born and where we live.

If, however, we take a clear-eyed look at our recent discoveries, we realize that many of the ideas of the past are being eroded, the first and most fundamental of these being the explanation of how the planets were formed. This is normal in science: we advance a step at a time and the most inspiring moments are linked to those new discoveries which help us to move forward, abandoning former beliefs. This is particularly true of the Solar System where, until relatively recently, everything seemed to have been clarified, explained and accepted as fact.

But just as the explorers of the sixteenth century unveiled the unknown face of the Earth, space probes have revealed a planetary system of which the greater part has yet to be discovered. After thirty years of exploration with cosmic robots we have managed to put together some interesting pictures of the planets, reaching out as far as Neptune, the most recent to be approached by the *Voyager 2* probe. Pluto, however, we still know only as a cloudy disk with surface markings, as photographed by the Hubble Space Telescope.

From work achieved to date, we should, according to the California Institute of Technology's distinguished planetologist David J. Stevenson, learn five lessons. The first demonstrates that similar processes, common to one another, are not only already taking place now but will continue to happen. The second is that these common processes often have different outcomes. The third lesson tells us that we must bear in mind the reciprocal influences that the bodies exert on each other and which help to determine what they are like. The fourth, that we must take account of history or, to put it another way, past evolution. Finally, the fifth lesson confirms that in order to exploit space exploration and the impressive images transmitted back to Earth to the fullest, all the research work and computer simulations carried out by planetary scientists on our home planet are still vitally important. This is truer than ever today, when science is moving inexorably towards a different planetology (the study of the planets) in which the work of comparison between the various bodies in space is of paramount importance if we are to discover the common and diverse elements that will provide an explanation for the system's origins, historical evolution and the possible future. This is why planetary scientists are studying the Great Red Spot of Jupiter in order to decipher the enigmas of terrestrial hurricanes, and why they are analyzing the greenhouse effect on Venus to understand the threat of global warming on Earth.

The scope of our potential fields of investigation has now opened up another perspective that adds to our fascination with the planets. This is the search for life which could be hidden under the ice of the moon Europa or in the depths of the Martian ground. Perhaps some primordial traces may be floating in the methane lakes of Titan, the natural satellite of Saturn. Or perhaps life once existed but has already become extinct on the Red Planet, after the water, once so plentiful in the deep canyons and in the wide oceans, had vaporized or sunk into the depths of the planet. If a fossilized micro-organism were to be discovered, it could tell us a story that would revolutionize our conception of history and change all our ideas.

Meanwhile another field of research has opened up: the exploration of that populous world of the asteroids and comets, the small bodies of the Solar System which were neglected until a few years ago but which have now acquired a higher priority, not least because of the threat which some of these bodies appear to pose to the Earth. Increasingly powerful observational instruments are revealing them in staggering, ever-increasing numbers. In addition to the asteroids traveling between the planets, most of which are found between Mars and Jupiter, we are discovering that a new family of trans-Neptunian objects exists beyond Neptune's orbit which means that we are even being forced to re-draw the already uncertain boundaries of the Solar System. This may well lead, according to various astronomers, to the deletion of Pluto from the list of major planets and to its relegation to the category of one of the trans-Neptunian objects.

There now exists, moreover, a particularly strong motivation to undertake this exploration of asteroids and comets, however close or far off they may be, and whilst the distinction between the two is becoming more blurred in certain cases. This field of exploration is becoming increasingly enthralling because these "minor bodies" as they have historically been called by astronomers, represent the best uncontaminated evidence of the origins of the Solar System. We are now able to investigate them by employing the new, often diminutive cosmic robots which are now intelligent and capable of replacing humans in places that mankind cannot reach. This is why we should have this little guide to the Solar System always at hand, to remind us where we came from and enable us to dream of our future.

Giovanni Caprara

Symbols

 Solar System

 Mercury

 Sun

 Venus

 Comets

 Earth

 Asteroids

 Moon

 Meteorites

 Mars

 Jupiter

 Saturn

 Uranus

 Neptune

 Pluto

THE SOLAR SYSTEM AND THE SUN

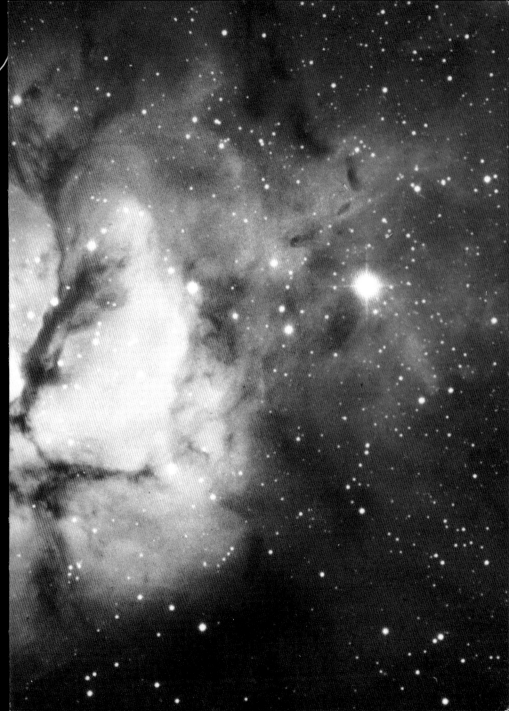

THE SOLAR SYSTEM

Above: The Orion nebula. Dark nebula and bright nebula are thought to be the nurseries where the young protostars are formed. The difference between dark and bright nebulae depends only on the fact that, in the latter, the gas of the clouds reflects, or is excited by, the radiation of very luminous stars. These clouds are concentrations of interstellar gas and dust, with a density ranging from 10 to 10,000 times higher than that of the diffuse interstellar gas.

One of the most difficult and controversial fields of astronomy is the study of the way in which the Solar System was formed, as well as its past and its future evolution.

The Solar System or, more particularly, its larger bodies—all the planets with their satellites, orbit in the Sun close to one plane known as the ecliptic which is defined as the plane in which the Earth orbits, and all move in the same direction. It is helpful to remember that an observer viewing this from a position above the plane of the ecliptic, or from the direction of the terrestrial north pole, would see the planets move around the Sun in an counter-clockwise direction, also referred to as prograde motion. All the planets and their major satellites rotate in an counter-clockwise direction on their axis, with the exception of Venus and Uranus which, because of the position of their axes shows that they experienced some cataclysmic event in their far-off past.

The distance of the planets and their satellites from the Sun increases regularly as described by the Titius-Bode law. This regularity in the movements of revolution and rotation, and the disk shape of the whole system suggested a hypothesis to Immanuel Kant as long ago as 1755, and to Pierre-Simon de Laplace in 1796: that the Sun and the planets had been formed out of a cloud of rotating gas as a result of the combined action of the force of gravity and centrifugal force. This cloud has historically always been known as the "solar nebula."

Another aspect that must be borne in mind is the distribution of density within the system: the terrestrial planets have higher densities, varying from approximately 5.5 and 3.9 times the density of water, while the giants planets have consistently lower densities, varying from 0.7 to 1.6 times the density of water.

The Age of the Solar System

Based on the age of the oldest meteorites, it has been calculated that the Solar System was formed approximately 4.5 billion years ago. Chondrules, spheres of crystal present in chondrite-type meteorites, contain radioactive elements and the proportions of these enable us to measure the time that has elapsed from the moment in which the crystals solidified: this time lapse provides us with an indication of the age of the Solar Sytem.

A group of research scientists at the Institute of Physical Earth Sciences in Paris has carried out very precise measurements of the relationship that exists in certain phosphates between radioactive uranium-238 and the element which derives from it, lead 206. As these crystals are rich in uranium but have virtually no lead in their natural state, the lead isotope present in the phosphates has to result from the decay of the uranium. Working from these measurements, radiometric dating means that we can calculate the age of the crystals at about 4.56 billion years.

The Sun and the Solar System

Given the close resemblance in physical structure and chemical composition of the stars and particularly of those of solar type, it is a reasonable

Above: An area of the Orion nebula, a gigantic cloud of gas, illuminated by young and very luminous stars, which is thought to contain several planetary systems in the process of formation.

Below: The position and distance of the inner and outer planets in relation to the Sun.

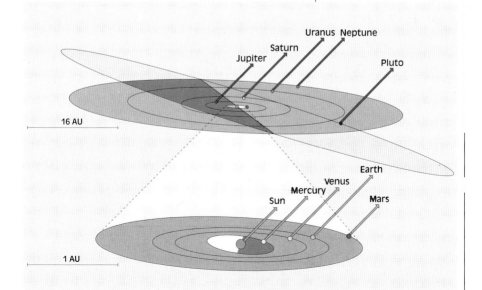

assumption that the Sun was formed in the same way as stars of later generations, those whose formation present-day observational techniques make it possible to witness. By the description "stars of the solar type" is meant those stars which have approximately the same mass as the Sun. The main physical features of a star and its evolution depend on its mass, as do the phases of its life: formation through the contraction of a cloud of dust and gas; maturity, when the star emits constant radiation for a length of time: the smaller the star, the longer the duration; and the star's end, with variations in radius and temperature and phenomena which may be more or less violent. It would appear that stars form from clouds of gas and dust, those of recent formation still being surrounded by these. The subsequent stages of stellar evolution are sufficiently well-known to make it possible to calculate their ages to within a very close approximation.

How a star is formed

Of all the phases of a star's life, we know least about its formation. These clouds have very low temperatures and are very opaque; as a result they

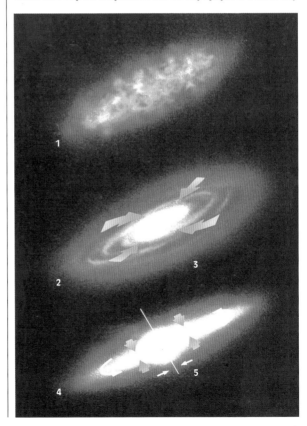

Right: A diagrammatic illustration of how a protoplanetary disk is formed by the collapse of a cloud of gas and dust: (1) when the cloud collapses, (2) gas and dust are attracted towards the center of mass of the cloud (3). Subsequently a rapid collapse towards the central plane occurs (4) and a slow collapse towards the cloud's rotational axis (5).

radiate only in the far-infrared which the terrestrial atmosphere absorbs completely, preventing us from seeing what is happening inside the cloud. As a result of observations from space using the *IRAS (Infrared Astronomical Satellite)* and data from microwave radio-astronomy which can even penetrate the opaque dust clouds, it has been possible to witness the first phases of condensation of protostars from the interstellar medium. This has led to the discovery that the protostar contracts and the matter accretes onto its equatorial region, while at the same time it expels matter violently along the two opposing directions of its polar axis. In this way a disk forms on the equatorial plane. The presence of a disk of solid matter in stars that have already formed was established by the same satellite but it is also possible to discern this from Earth if the image of the star is occulted with an opaque disk, so that the weak luminosity of the disk is not overwhelmed by the predominant stellar brilliance. These observations enable us to conclude that the young Sun must also have been surrounded by a disk of gas and dust: the solar nebula hypothesized by Kant and Laplace two centuries ago.

Below: NASA's Hubble Space Telescope, launched by the space shuttle Discovery (STS-31) in April 1990 facilitates reconnaissance of the planets in the Solar System. Significant results have been achieved in these fields, especially with observation of Mars, Jupiter and Saturn.

According to stellar evolutionary theories, the Sun took approximately 50 million years to reach the phase of stability in its evolution, that is to say, to reach its present values of radius and surface temperature and to radiate, at any given moment, the same quantity of energy that it emits today. Based on these same theories, it has been estimated that these values will remain virtually unchanged for another 5 billion years.

Composition of the stellar nebula

Dust represents only 2% of the stellar nebula mass; the remainder is made up in the following proportions: gas, of which 78% is accounted for by hydrogen atoms, 20% by helium atoms and 2% by all the other elements. The dust accounts for the largest percentage of the heavier elements which

Above: An illustration of the various phases in the formation of a solar system. Working from left to right, this begins with the contraction of the nebula disk (1); followed by the unstable phase in which the rings are formed (2); subsequently gassy protoplanets are formed from the rings (3).

Right: The disk of matter surrounding the star Beta Pictoris in an image obtained by occulting the image of the star with an opaque disk, which prevents the weak light of the disk from being completely obliterated by the stellar luminosity. It is thought that disks such as this are typical of a protoplanetary system from which the planets subsequently condense.

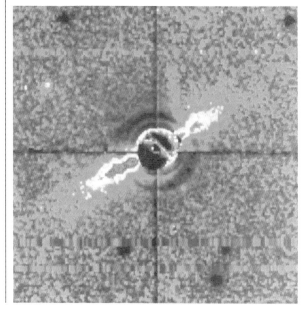

are present in planetary conformation: carbon, oxygen, iron, silicon, magnesium. The denser clouds contain between 10,000 and one million molecules per cubic centimeter and have temperatures of approximately 10 degrees Kelvin, the equivalent of 263°C (440°F) below zero. They are unstable because at such low temperatures the force of gravity is greater than the thermal pressure exerted by the gas and dust particles.

Origins

Of the numerous theories put forward to explain how the planets were formed, only two are still considered acceptable, although both of them pose many unsolved problems: the first is the planetesimal or accumulation theory, and the second is the protoplanetary or unstable disk theory. How can the formation of a planet from a disk of dust with particles measured in micrometers be explained? It is worth looking at the two possible answers.

Below: Other phases in the formation of a solar system. From left to right: a thin disk of dust (4) forms in the protoplanetary gas, and then small bodies of similar sizes to those of asteroids; in this phase the Sun's radiation causes the gas to disperse (5); finally, the protoplanets capture minor bodies, increasing their own mass (6).

Formation through accretion

This is the most widely accepted hypothesis to be formulated to date. The particles collided with each other and stuck together, leading to processes known as "coagulation" and, subsequently, "accumulation" or "accretion." In this way larger bodies grew from the granules which collected into yet bigger, solid bodies, until they formed terrestrial planets and the nuclei of the giant planets. The gases and dust remained closely intermingled for as long as the turbulent movements typical of the interstellar medium continued in the solar nebula. As the turbulence in the solar nebula diminished, the particles began a process of sedimentation, accumulating towards the central plane of the nebula, forming a very thin disk. Given the relatively high density, it is estimated that the sedimentation process was very quick, perhaps as fast as a thousand years. As the sedimentation process progressed, the mass of the thin disk became unstable and the disk broke up into a large number of solid bodies, each having a diameter of approximately 1 kilometer (.62 miles), called planetesimals. It is estimated that approximately a thousand billion of these existed, orbiting around the Sun, within the region encompassing the orbit of Mars.

The accretion of the planetesimals, as a result of random collisions, placed them on orbits that were very close to one another, favoring further accretion into a single body. It has been calculated that this process took approximately 10,000 years and that it produced planetesimals with radii of approximately 500 km (310 miles) which can also be described as "embryonic planets." As subsequent collisions occurred, once-separate embryonic planets were combined and finally formed the

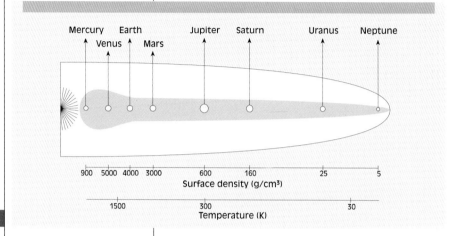

Above: This diagram shows the probable temperatures and densities of the nebular disk of dust during the first phases in the formation of the Solar System.

planets. It is thought that the terrestrial planets must have taken 10 to 100 million years to form. One of the greatest problems with the accretion theory is how to account for the formation of the giant planets.

Astronomical observations indicate that the nebula around the young, solar-type stars evaporate into interstellar space over periods varying from 100,000 to 10 million years after the formation of the star. It therefore follows that Jupiter and Saturn must have been formed first of all in order to have been capable of retaining fractions of hydrogen and helium in proportions comparable with those of the interstellar medium and of the Sun. The theory does, however, allow for a much longer period of time for the formation of the giant planets: approximately 100 million years or more. If this is the case, accretion could have happened at such a rapid rate that bodies with a mass equivalent to 10 times that of the Earth were formed within a period of a million years. Such large bodies could then have quickly captured hydrogen and helium, through gravitational attraction, in sufficiently large quantities to account for the mantles which surround the nuclei of these planets.

Instability of the disk

The second theory, an alternative to the accretion theory, postulates the formation of the planets as a result of the instability of the gaseous disk. There are, however, even more problems inherent in this theory than in

Phenomena that modify the crust of terrestrial planets

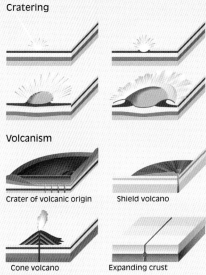

Cratering

Volcanism

Crater of volcanic origin | Shield volcano

Cone volcano | Expanding crust

Tectonics

Thrust fault | Strike slip fault | Normal fault

Atmospheres

Sunlight

Release of volatile gases | Emission of gases | Chemical reaction | Evaporation and condensation

Morphogenesis

Gravitational sliding | Wind erosion, transport and deposition | Fluvial erosion

Left: A diagrammatic illustration of the various phenomena that change the crust of terrestrial planets: these include impact cratering, volcanism, tectonics, and erosion by water and wind.

Atmosphere
Crust
Mantle
Core

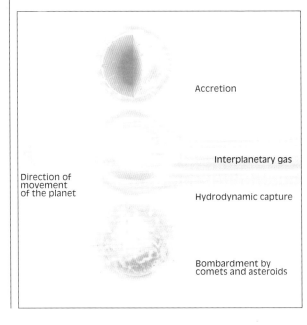

Right: A diagrammatic representation of the three main hypotheses as to how planetary atmospheres were formed. In the first (top) the volatile elements already present in the dust particles are freed when the planet is formed through the effect of internal heating. In the second (middle) the particles did not contain volatile elements but, once the planet had formed, it captured them from the primitive nebula. In the third (bottom) the planet has acquired its atmosphere after an intense bombardment by bodies rich in volatile substances from other parts of the Solar System and by asteroids and comets. The first hypothesis now seems the most plausible, although it does not preclude a small contribution from the volatile substances contained in the asteroids and comets (as postulated by the third hypothesis).

Accretion

Direction of movement of the planet

Interplanetary gas

Hydrodynamic capture

Bombardment by comets and asteroids

the preceding one. For the solar nebula to have had a sufficiently large mass, it would have to have been unstable and must have fragmented within a short time, creating large, gassy planets within the space of less than 10 years. For this process to have been feasible, the original nebula must have had a far greater mass than that of the present Solar System. It is difficult to explain how this excess mass could have been expelled: neither tidal forces nor thermal evaporation caused by solar radiation are sufficiently efficient to have brought this about.

Critical points hypothesis

An explanation has been put forward by A.G.W. Cameron. Between two bodies there is always a critical point at which their respective gravitational attractions are equal and opposite. Between the Earth and the Moon this point occurs at a distance of 38,397 km (23,860 miles) from the Moon and 345,570 km (214,737 miles) from the Earth, that is, respectively, at one-tenth and nine-tenths of the Earth-Moon distance.

These critical points, existing between the protosun and the protoplanets, must initially have been situated outside the planets themselves, but as the solar mass grew, and as the nebula collapse gradually took place, the protoplanets would have drawn ever closer to the Sun, until the critical point came to be located within their atmospheres. The external layers would thus have been stripped from the planet and have fallen onto the Sun. A process of this type would have been much more obvious on the inner planets, not least because the solar heat would have caused the greater part, if not all, of their atmospheres to evaporate rapidly. The outer planets, however, would not have come so close to the Sun and they would therefore have retained most of their original content. In addition, during the first period of the Sun's life, it certainly passed through a phase known as T Tauri, from the name of the young, newly-formed stars which can be observed today. During this phase it emitted a great quantity of matter at a very high speed. Thus the recently-formed Sun produced a strong solar wind which expelled most of the gases present in the nebula and in the atmospheres of the closest planets, while being unable to sweep away the dust.

Testing the theories

Several scientists have attempted to test these theories through computer simulation. Stephen H. Dole is one of them, and he has tried to test Cameron's theory. Even when starting from differing initial conditions, the results always lead to planetary systems with fundamental features which they have in common and which are similar to those of the Solar System: that is, that they consist of some small, terrestrial-type planets closest to the Sun, three or four large, low-density planets in the outer regions of the System and, finally, one or two small bodies on the extreme periphery.

Wetherill's research

More recently George W. Wetherill has carried out a series of computer simulations to find out what happens when interplanetary matter

Facing page, top: An imaginary portrayal of a scene on primitive Earth, its sky riven by lightning flashes producing ultraviolet light while the Moon, which is believed to have been far closer to Earth, looks enormous above the horizon.

Above: Under the action of the Sun's ultraviolet rays organic molecules and water gave rise to amino acids, the main constituents of proteins. These were probably the first steps towards the birth of life on primordial Earth.

Above: The planets of the Solar System are compared with the Sun to show relative sizes. From top to bottom: Pluto, Neptune, Uranus, Saturn, Jupiter, Mars, Earth, Venus, Mercury and the Sun.

Right: When the Sun is twice as old as it is now (its present age being approximately 5 billion years), it will turn into a giant red star and Earth's orbit will be swallowed up by its surface which could stretch out far enough to impinge upon the orbit of Mars.

accretes, taking into account gravitational disturbances and collisions in the region between the Sun and the asteroid belt.

The result of these simulations would seem to indicate that in the presence of a star like the sun, surrounded by a nebula of gas and dust, and having mass, dimensions and rotational speed which do not differ too greatly from those of the solar nebula, it is very probable that a planetary system will be formed, with internal planets distributed more or less as in the present Solar System.

Wetherill then went on to carry out many other simulations starting from slightly different initial conditions. He has made a particular study of the effect of Jupiter on the formation of the Solar System. Although Jupiter is responsible for grouping the asteroids along certain orbits, making collision and accretion more likely, Wetherill demonstrated through further simulations that even if Jupiter were not present, in three-quarters of cases a planet of the terrestrial type would be formed at a distance of approximately 1 astronomical unit. In the region that lies within 2 astronomical units there are often four terrestrial planets. The extraordinary thing is that despite the chaotic and random origin of the accretion process, this should lead naturally to a regular distribution of planets.

From this it can be deduced that the formation of planets similar to Earth is fairly frequent in the Universe and that the formation of a solar system is also a process which naturally accompanies the formation of a star, provided that the nebula from which it condenses has adequate mass and rotational velocity. To sum up, certain points are already widely accepted and these are: the age of the Solar System, and its formation from a rotating nebula collapsing under the action of its own gravity.

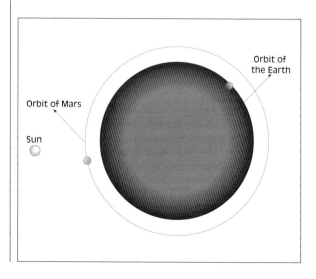

Left: The planetary nebula NGC 7293, also known as the Helix Nebula, in the constellation of Aquarius. The star in the center is all that remains of what was once a giant red star. Its rarefied envelope has gradually become detached from the central core, giving rise to a "shell" of gas which constitutes the planetary nebula, while the hot, dense central core is called a white dwarf.

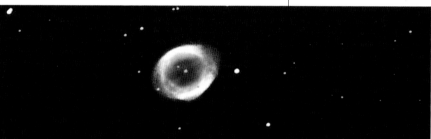

The Dying Sun

The future of the Solar System is closely linked to the Sun's future. The theories of stellar evolution and observations of the galactic stellar populations confirm that the Sun will remain more or less unchanged for another 5 billion years. A fairly rapid evolutionary phase will then begin, which will spell the end of the terrestrial planets. The Sun will undergo a transformation into a giant red star, with a radius 100 or 200 times its present radius, swallowing up within its surface Mercury, Venus, Earth and perhaps Mars as well. Its luminosity will become 100 times greater and the giant planets will also undergo dramatic transformations as a result of this, as they will be continuously buffeted by a solar wind that will be many times more intense than at present. After approximately half a million years, the extensive and rarefied atmosphere of the Sun will have completely dispersed into interstellar space and only the central core, small and hot, but 10,000 times less luminous than our present Sun, will remain. Bereft of energy sources, the Solar System, or what will remain of it, will gradually approach a slow end through cooling, together with its dying Sun.

Above: Another planetary nebula, known as the Ring Nebula, in the constellation of Lyra.

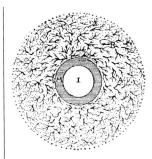

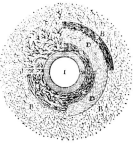

Top: Descartes depicts the first stage of increase in density of what was to become the Earth. In the central nucleus (I) is matter of the first element, in the middle region (M) there is opaque, dark matter, in the outer zone (A) there are particles of the third element which will produce the various materials on the Earth's surface.

Above: The progressive "thickening" of the particles of the third element from which the various solid layers or "envelopes" of the Earth and the atmosphere which surrounds it have originated, according to the Cartesian conception.

Right: The progressive loss of a fluid internal section led, according to Descartes, to the fracturing of the surface layers and the formation of mountain ranges.

The Original Nebula

In his Principia philosophiae *and his universe of the vortices, Descartes was the first person to draw up a scientific theory of cosmogony that allows for an irreversible evolution from the simple to the complex and links what happens to the universe to the immutability of his system's laws. At the beginning of things, Descartes places only space endowed with motion. With this undifferentiated and primordial* res extensa *he describes the diversity of the smallest particles and the vortices through which they are drawn across the skies. In the vortices, according to Descartes, the finest and lightest particles move away from the center towards the exterior, while the larger and heavier particles descend towards the innermost zones where they tend to aggregate, forming the heavenly bodies.*

A second mechanistic hypothesis was promoted by Georges Louis Leclerc, Marquis de Buffon, the great eighteenth-century naturalist. He put forward the hypothesis that the planets could be derived from a tongue of matter torn away from the Sun as a result of a comet or another star passing close by, this matter then broke up and evolved into the planetary bodies, which were originally liquid, from which the satellites were, in turn, also derived.

The first person to revisit the Cartesian theory from the perspective of Newtonian physics was Immanuel Kant. He postulated the initial existence of matter, uniformly distributed throughout space and immobile, but to which the elements, possessing their own forces, gave shape. According to Kant it is the force of gravitational attraction that increases the density of matter and endows it with motion, giving rise to the first nebulae and the condensations within them. Such motion had progressively to assume some sort of order, until they were confined to circulation around the largest condensation, which would become the Sun. The orbits acquired in this way would all be circular, coplanar (in the same plane) and be described in a single direction,

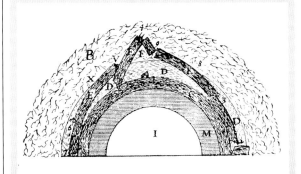

otherwise the various condensations would disturb neighboring condensations. Once this position had been reached, the process of condensation would continue, leading to the formation of the planetary bodies and the Sun.

In his Exposition du système du monde *of 1796, Pierre-Simon de Laplace, the great French physicist and mathematician, put forward the hypothesis of a "primordial nebula," which differed from the Cartesian vortex because it did not evolve solely as a result of the action of inertia, but through the Newtonian force of attraction. According to Kant, the planets originated from condensations inside the nebula; Laplace maintained that they originated with rings of material expelled by the nebula during the process of contraction which led to the formation of the Sun through the continual increase of centrifugal force. Friction inside the nebular vortex, or the form of motion of the rings, ensure that the planets' orbits are coplanar, substantially circular and move in a common direction. The satellites are derived from condensation of the nebulae and the protoplanetary vortices, similar to the theories advanced by Descartes. Despite its limits, the primordial nebula described by Kant and Laplace provided a qualitative explanation for much of the regularity of the disposition and movements of the bodies in the planetary system, and put forward reasons why all the planets experience their coplanar, low-eccentricity orbits in the same direction which is, moreover, the direction of axial rotation for the planets, the Sun, and the satellites around the planets. This theory was very widely accepted, but only in recent times has been possible to re-assess it critically from the perspective of present-day physics and astronomy.*

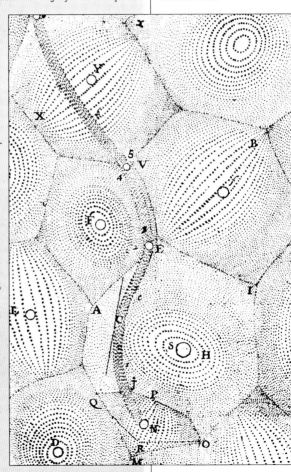

Below: The universe of vortices according to Descartes. S stands for the Sun, surrounded by other vortices, with other suns at their centers. The winding trail which starts below at N stands for the path taken by a comet moving from vortex to vortex.

THE SUN

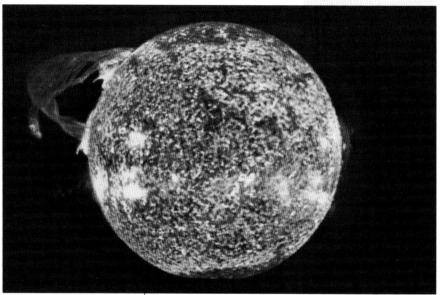

Above: A gigantic prominence recorded by Skylab in the light of ionized helium. This was one of the largest prominences observed over a period of ten years. Its temperature is approximately 20,000 K, but some more intense striations reach 70,000 K. The streamers of gas have a spiral shape: this is an indication of the presence of magnetic forces.

Physical characteristics

Among the 300 billion and more stars that form the Milky Way, the Sun is one of the most common; it could be described as one of its "typical citizens." In fact, bearing in mind that mass is the main feature that defines the physical properties of a star, the Sun is half-way between the relatively few, very bright stars of greater mass, ranging from approximately 20 to 100 times the solar mass, and those of lesser mass which range from approximately one-tenth to one-twentieth of the solar mass and which are very numerous and very faint. In order to describe the physical structure and properties of the Sun it is necessary to start with a description of its various component parts: the central core, where the nuclear reactions take place, the source of the energy that the Sun has been irradiating at a constant rate for over 4 billion years; the interior; the visible layer or photosphere; and, finally, its atmosphere, consisting of the chromosphere and the more rarefied corona. The next step is to analyze the sunspots and the prominences, the flares, the hotter coronal zones, and the coronal holes: that is to say, every aspect of the activity that occurs in its photosphere and its atmosphere, the only regions that can be directly observed. In common with all stars, the Sun is a completely gaseous sphere, its temperature decreasing from the core, where it reaches 15 million degrees Kelvin (K), to the photosphere, where the temperature is approximately 6,000 K.

The pressure at the center is equivalent to approximately 200 billion atmospheres, falling to a tenth of atmospheric pressure near the photosphere.

A balance of opposing forces

The Sun is not collapsing inwards, into its own center, nor is it dispersing into interstellar space because in it two equal and opposite forces are working against each other: the force of the pressure exerted by the particles of hot gas that proceed outward in a random walk, and the force of gravity which, if unopposed, would make the entire solar mass collapse inwards, towards the center of the Sun. Because of the high temperatures that exist on the Sun and, generally speaking, on all stars, the gas of which they consist is wholly, or to a large extent, ionized. This means

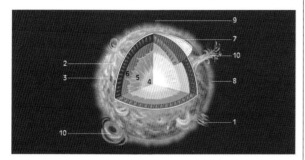

Internal structure

1—filament
2—facula
3—sunspot
4—core
5—radiative zone
6—convective zone
7—photosphere
8—chromosphere
9—corona
10—prominence

Characteristics of the Sun

Absolute Magnitude	+4.83
Apparent Magnitude	−26.74
Surface Temperature	5,770 K
Core Temperature	15.3×10^6 K
Luminosity (Watt)	3.83×10^{26} K
Spectral Type	G2 V
Mass (g)	1.989×10^{33}
Radius (km)	6.960×10^5
Mean Density (g/cm^3)	1.410
Core Density (g/cm^3)	140–180
Solar Constant (watt/cm^2)	0.137

that the atoms have lost one or more electrons and are electrically charged. We cannot observe the innermost regions of the Sun, but only the surface and the more rarefied atmospheric layers. Where the properties of the Sun's interior are concerned, we have to resort to an indirect approach, through theoretical proof and computer simulation, using a

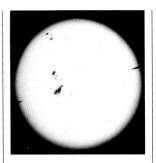

Top: The Sun as seen through a telescope. Groups of sunspots and small, isolated sunspots are present. Sunspots appear dark because they are cooler than the rest of the solar surface: approximately 4,300 K as compared with 6,000 K. On the limb, which is less bright than the center, faculae (hotter regions) can be seen.

Above: A detail of the large sunspot group visible on the solar disk above. The umbra and the penumbra are clearly visible, as are numerous smaller sunspots between the two largest ones, and granulation.

model that must include those features that can be subjected to direct measurement—for example, the surface temperature and the quantity of radiation emitted.

The solar surface

The photosphere, or "luminous sphere," is the name used to denote the solar "surface" that looks like a clearly delineated sphere, almost as if it were a solid globe. If, however, the Sun is observed through a telescope, even an unsophisticated one, it can be seen that the edge is slightly blurred, a sign of the semi-transparency of the most superficial layers. In addition, the solar disk is not uniformly bright: its brightness decreases slightly towards the "edge" or *limb*. This phenomenon, known as "limb darkening" is due to the fact that, when looking at the center of the disk, light radiation from layers situated along the solar radius reach the observer, to a depth of approximately 600 km (400 miles).

The light radiation coming from deeper layers is practically all absorbed by those above them. When, however, we look at the limb, the radiation still emanates from layers extending to a depth of approximately 600 km (400 miles), but these are sited tangentially in relation to the solar surface and are therefore, on average, colder and have lower emissivity than those which contribute to emission at the center of the disk. When the Sun is viewed through a telescope it is possible to make out other features: sunspots, faculae and granules.

Sunspots and faculae

Sunspots appear as dark patches, isolated or in groups; faculae are lighter areas, mainly situated around the groups of sunspots or isolated, on the limb of the solar disk.

The central, darker part of the sun spots is known as the umbra and is surrounded by a grayish penumbra. The sunspots look dark because their temperature is approximately 1,000-1,500 K lower than that of the surrounding photosphere. Were it possible to observe sunspots in isolation, they would look bright.

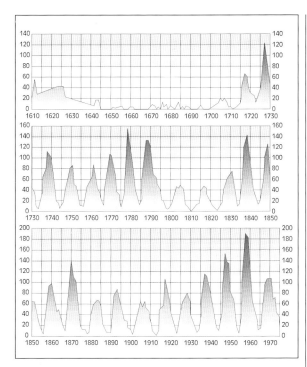

Left: The number and frequency of sunspots increases and decreases periodically. Approximately every eleven years a maximum of sunspots occurs. The first observations of these date back to 1610, recorded by Galileo. Maxima and minima vary from one cycle to another. It is noteworthy that in the period lasting from 1640 to 1710 maxima were practically non-existent. Sunspots are an easily-observable indication, even with unsophisticated telescopes, of the level of solar activity. The variability of the number of sunspots was noted immediately after their discovery, but the cyclical pattern, with an average interval of 11.2 years between two successive maxima (or two successive minima) was only discovered in 1843, when R. Wolf, working in Zurich, Switzerland, carried out detailed analysis of documentary records of solar observations dating back to 1610. As a result, he defined what were subsequently known as the Wolf numbers, still used today to indicate sunspot frequency. This number, indicated by R, is obtained by multiplying the number of groups by 10 and adding it to the total number of sunspots observed each day. The numbers shown in the ordinate of the graph are the average Wolf numbers for each year.

Facing page, below: A spectacular photograph taken by Mark Cunningham in Colorado during the solar storm of March 2001.

An important feature of sunspots is their magnetic field which forces the ionized gas to spread out along the force lines of the field, in exactly the same way as iron filings arrange themselves between the two poles of a magnet. The larger the area of the sunspot, the greater is the intensity of the magnetic fields inside them. Magnetic fields can therefore measure anything from approximately one hundred to several thousand gauss, while the general magnetic field of the Sun is approximately 1 gauss. Consequently a sunspot is like a gigantic electromagnet.

Sunspots were discovered and systematically observed by Galileo from 1610 onwards and they provide us with an index of solar activity, easily seen through ordinary telescopes. Although the radiation emitted by the Sun is virtually constant and those variations which may occur are too small to be measured, there are many phenomena that signal a change in the level of solar activity; despite the fact that these can also be very spectacular, they nevertheless have a negligible effect on the sum total of solar energy. A colossal sunspot, 13 times the surface area of the Earth, faced our planet at the end of March 2001, unleashing the most violent solar magnetic storm of the decade. This reached its culmination on March 29 with an eruption, the effects of which were soon felt by the terrestrial magnetic field, in the form of spectacular *aurorae* and interference with telecommunications.

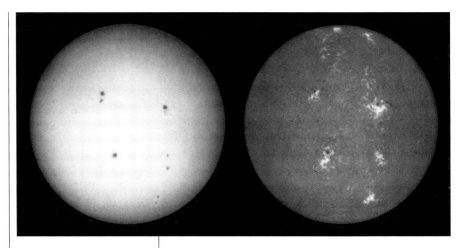

Above: Two simultaneous images of the Sun. Left: in white light; right: in the light of the most intense emission of hydrogen, the most abundant element of the Sun (and, generally, throughout our universe), known as Hα, yielding a splendid bright red. The matching sites of sunspots and brighter areas in Hα light are obvious.

Below: A picture of the Sun in Hα light. The long, dark filaments are prominences projected onto the solar disk: they appear dark because their temperature is lower than that of the chromosphere. The brighter areas are hotter regions and these are usually associated with sunspot groups.

The lower radiation emitted by the sunspots is somehow balanced by the presence of areas of the photosphere that are somewhat hotter: the faculae, which, appear on the solar limb and near the sunspots. Observing that the sunspots appeared to move on the solar surface, Galileo inferred that the Sun rotates around its own axis and, from the elliptical path described by the spots, he even worked out that the solar equator is tilted in relation to the plane of the ecliptic.

Today we know the precise value of the solar radius and from measurement of these movements it has been concluded that the Sun's equatorial rotational velocity is 2 km/s (1¼ miles/s) and its equatorial period of rotation is approximately 25 days and 9 hours.

Distribution of sunspots

Sunspots almost always occur between two bands, parallel with the equator and between the latitudes of 5° north or south and 40° north or south. Observation of their apparent movement on the solar surface has revealed the fact that the Sun's angular speed of rotation is not the same at all latitudes, but decreases towards the poles: in other words, the Sun does not rotate like a rigid body. To illustrate this we can compare the wheel of a bicycle, which is obviously a rigid body and has the same angular velocity at each point of its radius, with the vortex or "whirlpool" of water that forms around the plug hole of a basin as it drains away, rotating at angular velocities that are greater near the center of the vortex and gradually decrease towards the outer edge.

Analogous to this, and as a result of the same phenomenon, the layers of the Sun situated at various latitudes "slide," in a manner of speaking, over one another, producing currents of ionized gas that not only influence and modify the magnetic fields but also have important effects on solar activity. The number of sunspots and the total area affected by groups of sunspots varies cyclically with time. The Swiss astronomer, R.

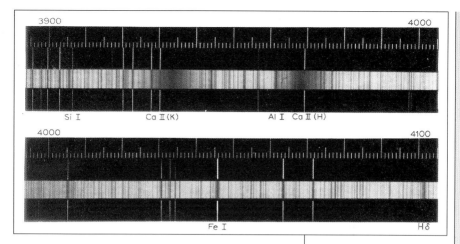

Wolf, amassed a collection of all the documented observations of sunspots, from the time of their discovery by Galileo onwards and, using this data, demonstrated that a solar cycle existed, lasting approximately 11.2 years, which means that the Sun shows maximum activity on average every 11.2 years. The number that Wolf chose in order to represent solar activity and which is still in use is obtained by multiplying by 10 the number of sunspot groups and adding the result to the total number of individual sunspots observed each day, then calculating the average for every day of a given year.

Granulation

Another feature of the solar surface is its mottled appearance: small, light and dark areas (known as granulation) can be seen which are the changing tops of currents of, respectively, rising hotter gas and descending, cooler gas. The size of each granule varies between 600 km (400 miles) and 1,000 km (600 miles) and they subtend an angle of 1–2 arcseconds when seen from the Earth. The maximum difference in temperature between light and dark granules is 100–200 K. A helpful analogy for these granules is that of grains of rice being carried upwards and then downwards by the rising and falling currents just like rice being cooked in a pot of boiling water.

Analysis of the solar spectrum

The astronomer's most effective tool in the study of the physical structure and chemical composition of the Sun and the stars is the analysis of the radiation they emit, or, to put it more succinctly, spectroscopy, the study of the spectrum. When the light emitted by a light source is made to pass through a dispersive medium, such as a glass prism, for example, it produces a band of visible colors ranging from red to violet. In the case of the Sun and the stars, the spectrum is composed of a continuum emit-

Above: A Solar spectrum obtained by the solar tower telescope of the Mount Wilson observatory in California.

Below: A picture of the Sun taken with a filter that only admits the light of ionized calcium (which is lacking one of its electrons) in extreme ultraviolet, at the limit of human optical perception. It reveals the appearance of the higher regions of the solar atmosphere (the chromosphere).

Right: A schematic representation of the solar spectrum: only the stronger dark lines are shown. The continuous spectrum, which is a rainbow band of colors ranging from violet to red, is emitted by the photosphere and the dark lines correspond to the characteristic wave lengths of various elements that absorb the light of the photosphere. Comparing the position of the dark lines in this spectrum with the laboratory spectra of various elements, we can identify those elements responsible for absorption and, hence, the chemical composition of the Sun's photosphere.

Below: Pictured in ultraviolet light, an eruptive prominence ejected by an extremely active region visible on the Sun's limb; this prominence extends to a height of approximately half a solar radius.

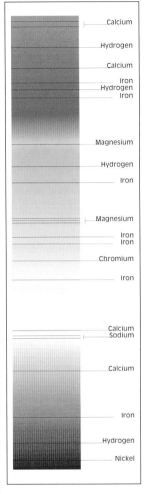

ted by the photosphere and interrupted by dark lines due to the various gases which occur in the uppermost areas of the photosphere and which absorb its light. The hotter the stellar photosphere, the more intense is the continuous spectrum at higher frequencies (short waves—towards blue). For example, the Sun reaches maximum emission in yellow-green. Colder stars, with surface temperatures of around 3,000 or 2,000 K, have a maximum emission in red and infrared, while hotter stars, from 10,000 to 30,000 K, have maximum emission in ultraviolet, only detectable from orbit. Bearing in mind that every element in its gaseous state emits its own, characteristic spectrum that identifies it in the same way as DNA identifies a living organism, these lines of the spectrum provide us with a great deal of information about the chemical composition, temperature and density of the surface layers; the movements of ascending and descending currents; the Sun's rotation, and the intensity of the magnetic field. Comparing spectra of various elements produced in their gaseous state in the laboratory with the spectra of the Sun and the stars, we can discover their composition and it is therefore possible to find out which elements are responsible for the lines that are present and carry out an initial, qualitative chemical analysis. Since, moreover, a given element is able to emit or absorb radiation only under certain conditions of temperature and density, we can determine the surface temperature and density from the presence and intensity of particular lines.

Doppler effect

When the position of the lines in the solar spectrum is slightly different from that observed in the laboratory, it is usually due to the well-known Doppler effect: if a source of light or sound waves is traveling towards us, we perceive waves of greater frequency than those emitted by the

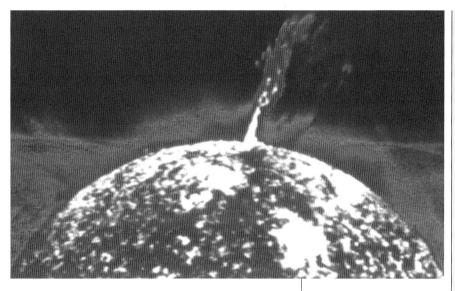

Above: A picture of the Sun taken by the space laboratory, Skylab, in ionized helium light at 304 Å, in extreme ultraviolet, inaccessible from Earth. The color yellow indicates the hottest, brightest areas.

source, and of lesser frequency when the sound source is receding away from us. It is common knowledge that the whistle of a train sounds higher when it is coming towards us, and lower if it is traveling away from us. When applied to light, the Doppler effect enables us to measure movements of the stars and also of the currents of gas on the Sun's surface. Understanding the Doppler effect has made it possible to measure the movement of convection currents: the dark granules emit lines that shift slightly towards red ("red-shift") whereas those which have originated at the greater depths of the photosolar limb that is moving towards the Earth as the Sun rotates: these exhibit "blue-shift," whereas those of the receding limb show "red-shift." This means that we can obtain confirmation that the Sun rotates on its own axis at an equatorial velocity of about 2 km/s (1.2 miles/s). It has also been observed that the equatorial rotational velocity of the photosphere is lower than that of the higher atmospheric layers.

Observation of the Solar atmosphere

The strongest lines by far in the solar spectrum absorb nearly all the radiation emitted by the photosphere. Astrophysicists have therefore devised a method that enables them to probe the solar atmosphere at various altitudes. When a colored filter that only allows the light of one of the strongest lines to pass through it is used to photograph the solar disk, this will result in a picture of the outermost part of the Sun, the chromosphere. Using filters with frequencies centered on other, weaker, lines, we can observe lower layers of the chromosphere, until (using no filters at all) an image of the photosphere in white light will appear.

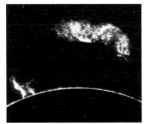

Above: A huge active prominence (i.e. changing within a short space of time) observed on May 29, 1919 in the light of the K line of ionized calcium in extreme violet, barely perceptible to the human eye. The image below was taken at 1.41 a.m. Greenwich Mean Time at Mount Wilson, in California, when it was still daytime. The image in the middle was taken at 2.57 a.m. and the one above it at 5.33 a.m. The prominence appears on the Sun's limb as an arch over 500,000 km (310,700 miles) high, which then rises above the limb until it becomes completely detached, reaching an altitude of just under one solar radius. This apparent challenge to the force of gravity is explained by the presence of the force lines of the Sun's magnetic field which supports the ionized (i.e. electrically charged) gas.

Attempting to observe the Sun through a strong, dark spectral line is like looking at the bottom of a receptacle full of cloudy water: all that can be seen is the surface of the water. If, however, the water is clear and transparent, it will be possible to make out details at the bottom.

A journey through the Sun

An imaginary traveler setting out from the center of the Sun and journeying from there to its outermost and most rarefied periphery, would pass through extremely changeable and varied surroundings.

At the center the traveler would encounter a powerhouse of hydrogen nuclei (or protons) and helium nuclei that move randomly in all directions and at very high velocity: these dynamics and their high density mean that the protons collide violently with one another, producing deuterons (or heavy hydrogen) and from them helium nuclei. The quantity of lost mass lacking in helium nuclei compared with the previous phases has been converted into energy in accordance with Einstein's well-known equation: $E = mc^2$, where c stands for the speed of light. Since the square of c is a very large number, even a tiny quantity of material can release very large quantities of energy. During the course of these nuclear fusion reactions, a large number of neutral particles are emitted of such small mass that it has not yet been possible to measure them in any satisfactory way. Because of such characteristics, these particles have been given the name of neutrinos and, again, because of the absence of electrical charge and the virtual absence of mass, they pass intact through the entire sun at a speed very close to the speed of light.

Temperatures and density only remain sufficiently high within a range of approximately two-tenths of the length of the solar radius to allow nuclear reactions to take place. A discontinuity is therefore created in the mass of gas: a central core that is gradually becoming richer in helium as it exhausts its hydrogen, and a mass of gas situated between the core and the surface through which the energy provided by the

nuclear reactions is propagated. Progressing outwards towards the surface, the temperature and density gradually diminish. All the energy produced in the core reaches the surface and radiates outward into interstellar space. But during its journey through the Sun, the energy changes in nature: at the center the highest energy photons are emitted, known as gamma-rays. As a result of the gas undergoing innumerable processes of absorption and re-emission, the original photons or gamma-rays are transformed into ever more numerous and less-energetic, X-ray photons, then into ultraviolet photons and, finally, into optical photons which, as their names indicates, are within the range of waves perceptible by the human eye. At approximately 200,000 km (120,000 miles) from the surface, currents of rising, hotter gas and of descending, cooler gas occur: in other words these layers of gas behave like water boiling in a saucepan and the tops of the currents can be seen in the form of the granulation described on page 31. These currents push upwards until they have nearly reached the surface and, as their density diminishes they change into shock waves that "slap" violently against the outermost layers, the chromosphere and the corona, heating the external layers. Put more succinctly, the energy produced in the center of the Sun is transported towards its exterior, initially by radiation (like light from any source, that irradiates and warms, and is also capable of traveling through a vacuum) and then by convection (hot masses that rise and cold masses that fall again, just like water boiling in a saucepan).

Above and below: A spectacular prominence recorded on July 24, 1999. The Earth has been added to this image to give an idea of the scale of this phenomenon.

Facing page, below: A picture of the Sun taken by the space laboratory, Skylab, in ionized helium light at 304 Å. A plume seems to spurt out of the Solar surface. The colors range from black, to dark red, to white, depending on the brightness of the area.

Right: A prominence, in the form of a closed arch, forming a loop, observed on the Sun's limb on August 3, 1970 in the light of the Hα line of hydrogen. The physical properties (temperature and density) are very similar those of the chromosphere where this prominence originated.

Below: A schematic representation of the force lines of the magnetic field of the solar corona; during periods of minimum activity (sunspot minima) the corona extends along the solar equator and only yields low prominences in the polar regions (above), while in periods of maximum activity, it is almost circularly symmetric (below).

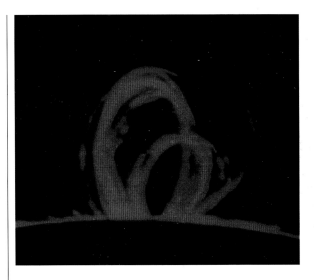

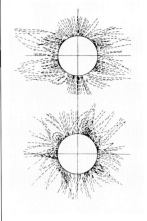

Analysis of the spectrum of the chromosphere and of the corona indicates that having reached a minimum of approximately 4,000 K at the edge of the photosphere, the temperature then goes on to rise to 10,000 K and subsequently to 20,000 K in the chromosphere, which consists of innumerable gassy flares called spicules, nearing an amazing one million degrees Kelvin in the corona. The term "temperature" does, however, need to be explained in this context: the chromosphere and the corona, unlike the photosphere, are very rarefied and transparent, and therefore by "temperature" is understood energy, or the velocity at which the atoms of the gasses move around. In the photosphere the density is equivalent to one millionth of the density of water, and then gradually decreases in the chromosphere, until it reaches a value of 10 million-billionths of the density of water in the transition zone; eventually its value falls by another 50 times in the corona.

Solar prominences

Immersed in the rarefied coronal gas, prominences develop. These are clouds of gas that have temperatures and densities comparable with those of the chromosphere and they are most frequently observed during periods of maximum activity, that is, at times when there are a greater number of sunspots. They can assume the most varied and strange shapes, such as arches, or fountain jets, and can be quiescent or active. The former hardly change their shape at all, even during the course of one or more solar rotations; the latter, however, change rapidly in shape and intensity, appearing and disappearing within the course of a few hours.

Solar flares

Another indication of solar activity at the chromospheric level is the frequency with which flares appear. These are regions no more extensive than the sunspots and they grow in brilliance within the space of a few minutes and then slowly revert to their original brightness. The whole phenomenon lasts for about ten minutes from start to finish. Flares usually occur between two sunspots of opposing magnetic polarities that form part of a large group.

The solar corona

Observation of the solar photosphere is undoubtedly the easiest way in which to study the Sun and dates back to 1610. But the faint solar corona, with a brightness comparable with that of sunlight diffused in the terrestrial atmosphere, can only be observed during the fleeting moments when a solar eclipse takes place. In fact, when observed from the Earth, the light of the corona is completely drowned by the brightness of the sky near the Sun. Only on a high mountain or on spacecraft, using a special instrument called a coronagraph which reduces scattering to a minimum, is it possible to observe the innermost and brightest area of the corona, using a filter that only allows light from the most intense emissions to pass through. Formerly, everything that was known about the corona had effectively been learned during total eclipses. These last for seven minutes at most (usually much less) and are only visible from Earth within a fairly narrow track of territory. Since our planet is three-quarters covered by oceans, it is very rare for a total eclipse to be seen

Below: The corona taken in white light, from the Skylab space laboratory. The false colors of this image show the differing temperatures of the corona. The background of a completely dark sky, excepting the terrestrial atmosphere, made it possible to carry out detailed observations of the outermost and most tenuous parts which extend to a distance of over 50 solar radii.

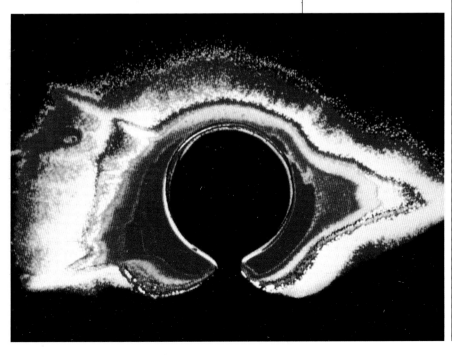

Right: X-ray photographs of the Sun, taken by Skylab, show a series of coronal loops that link areas of strong and opposing polarity on the solar surface. The magnetic fields generated in this way are much stronger than their terrestrial counterparts; the larger the sunspots, the stronger the fields.

Below: These three x-ray images, obtained with increasing exposures, show identical magnetic regions on the Sun, including the brighter parts (above), and the fainter parts (below).

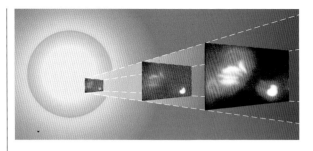

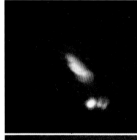

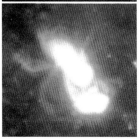

from places where there are astronomical observatories with all the necessary equipment. About thirty years ago, rockets and satellites with astronomical instruments on board started to explore the sky. Before the space age scientists were well aware that the shape of the solar corona varied according to the solar cycle. Total eclipses that took place during periods of maximum solar activity made it possible to see a corona that was almost circular, while in periods of minimum activity, eclipses enabled observers to see small prominences in the vicinity of the poles and to observe a corona that was very elongated in the direction of the solar equator. In periods of intermediate activity the corona also has an intermediate shape, often fairly asymmetrical.

Analysis of the coronal spectrum

Against the background of a weak continuous spectrum, similar in every respect to the solar spectrum and caused by diffusion of photospheric light by coronal particles, bright lines stood out, the strongest of which were a red line and a green line. The question of which elements were responsible for producing these coronal lines, remained unsolved for a long time. In fact, while comparison of spectra obtained in laboratories with the spectrum of the photosphere had led to the identification of lines caused by hydrogen, calcium and various metallic elements, among which were many lines yielded by iron, the spectrum of the chromosphere had revealed the presence of an element that could not at first be identified and was called helium, from the Greek name for the Sun, and was subsequently also detected on the Earth. In the case of the corona the problem proved to be far more complex: not one of the many coronal lines could be identified.

Theoretical developments in spectroscopy enabled scientists to calculate the spectra of elements that had undergone numerous ionization processes and to realize at last that the coronal lines were caused by elements that were very common but whose atoms had lost many of their electrons. This appeared to show that the coronal temperature was at least half-a-million degrees Kelvin, rising to more than one million degrees Kelvin. The strongest red line was yielded by iron lacking nine of its 26 electrons; the strongest green line was also yielded by iron, lacking as many as 13 electrons. The appearance of weaker lines, or the strengthening of weak lines already present coincided with the appearance of centers of solar activity: among these was the yellow line of calcium deprived of 14 of its 20 electrons. As to why the temperature of the corona is so high: this is still not at all clear, although these values have been confirmed by radio-astronomical observation which, with very different methods and instruments, confirms the increase in temperature from 20,000 K in the chromosphere to one million degrees Kelvin in the corona.

Discoveries about the corona in the space age

Observations carried out by various space laboratories and satellites orbiting around the Earth have meant that the solar corona can be subjected to continuous study, no longer confined to total eclipses, even in regions of the spectrum that cannot be examined from Earth, such as ultraviolet and X-rays, which are completely absorbed by the terrestrial atmosphere. In fact, the absence of the atmospheric shield eliminates the primary cause of scattering of sunlight; the sky appears completely black, while the Sun, the solar corona and the stars stand out clearly. In order to observe the corona in white light or in those regions of the spectrum in which emissions from the photosphere are present, it is necessary to hide or *occult* the image with an opaque disk, otherwise the intensity of light would damage the sensitive detectors used. If, however, the corona is observed by X-ray spectrometry, no screening is needed. In fact, the photosphere, with its temperature of 6,000 K is a very weak source of X-rays and the corona has turned out to be by far the most significant source. The *Skylab* space laboratory made it possible to take a great number of superb pictures of the corona. In these, as in those already taken from Earth in visible light, the presence of a magnetic field is obvious, distributing gas along its force lines.

Above: The solar corona as observed during the total eclipse on June 8, 1937 from the island of Canton, in the southern Pacific Ocean.

Below: An image of the solar corona that is typical of a phase of minimum activity. The picture is the result of combining a photograph in white light taken during a total eclipse and an X-ray image of the solar disk in false colors.

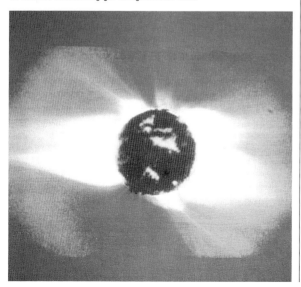

One of the main results obtained through observation from space relates to the structure of the corona. It is held in place by a framework consisting of magnetic structures, which can be likened to arches and loops, that support the highly ionized gas: in those places where these structures are not present, the corona is also absent and coronal holes occur. These are regions where both temperature and density are much lower and where the magnetic structures are not closed, but open, oriented, that is, in a radial direction towards interstellar space.

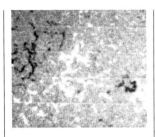

Above: This magnetogram reveals areas of opposing polarity on the Sun's surface in the form of light and dark areas.

Right: In this X-ray image of the corona, the different zones can be seen, with various features: the north pole (a) and an arm (b) that extends from it appear dark; these are the so-called "coronal holes" where X-ray emissions are lacking because the corona is missing. Thin, and more extensive loops (c, d, e) are present, linking regions of opposing magnetic polarity; bright regions or bright points (f, g) scattered over the entire surface; the corona on the limb (h) and at the south pole (i) and a dark coronal hole (j).

Right, below: An X-ray image of the Sun taken by Skylab. As is usual, the white regions are the hottest, occurring over active regions on the solar surface. In these regions the coronal temperature exceeds one million degrees. Because the chromosphere and the photosphere have much lower temperatures, they do not emit appreciable X-rays; thus what we see here is an image of the corona over the entire solar surface and not only those parts of it that can be seen during total eclipses.

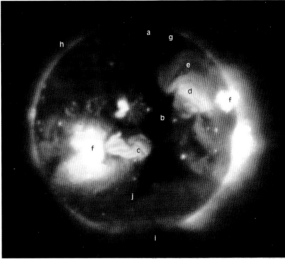

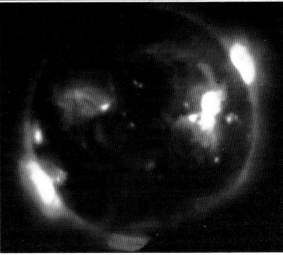

The corona is punctuated by small, bright points which emit the most energetic X-rays, and where temperatures rise from 3 to 5 million degrees Kelvin. This knowledge has been gained through recent discoveries and underlines the pivotal role of magnetic fields where hotter and cooler zones of the corona are concerned. The corona does not have a precise boundary: what ultraviolet and X-ray optical instruments enable us to observe is still part of the corona. Even during solar eclipses, what is visible from Earth of the corona stretches out to a distance from the Sun up to 10 solar radii.

Below: The solar corona photographed from East Africa during the total eclipse of June 30, 1973. The faint striations are reminiscent of iron filings between the two poles of a magnet, clearly showing the presence of the magnetic field. The shape is typical of a period of minimum activity.

Its outermost parts, however, merge gradually into the interplanetary medium: it is therefore thought that the corona extends even further and that only its weak luminosity prevents us from seeing its periphery. In white light, from airplanes at high altitudes and from space, the corona can be observed to a distance of over 50 solar radii. The luminosity of the innermost areas and the outermost zones of the corona is due to two distinct factors: the first mainly involves the scattering of photospheric light by the electrons of the solar atmosphere, while the second, for the most part, involves diffusion caused by minute solid particles present in interplanetary space. More recent measurements suggest it extends to 100 solar radii, equal to half the distance between the Earth and the Sun. It is possible that the Earth, and perhaps Mars also, orbit within the outermost layers of the corona.

The solar wind
Another feature of the corona, apart from its extension, is its continual expansion into interplanetary space. This constant flow of material has come to be known as the solar wind, an evocative description. This

Right: This is a sectional diagram of the interior of the Sun, from the core to the surface, showing the various parts of its atmosphere. In the central core the nuclear reactions occur, transforming hydrogen into helium, which is the source of the energy radiated by the Sun. Next comes a layer known as the radiative zone and, above it, the convective zone in which the energy is transported towards the exterior, by radiation and by convection.

Below: An X-ray image taken by Skylab. This shows very clearly an extensive coronal hole (the black area looking a little like the "boot" of Italy). On the left hand side of the image two large white patches can be seen which indicate more active coronal regions.

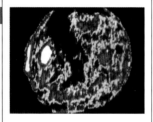

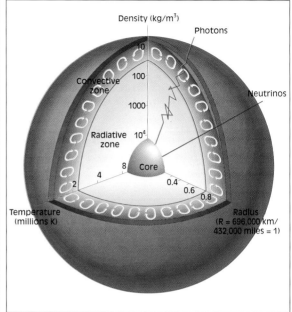

means that the Sun loses approximately one hundred-millionth-billionth of its mass every year.

This is a negligible quantity if we consider that the Sun was formed 5 billion years ago and that it will remain practically unchanged for another 5 billion years. In all this time it will therefore only have lost one ten-thousandth of its mass. It is, however, true that in the first few million years of its life, loss of matter was certainly much greater. The Sun is not, however, exceptional when compared with other known stars. Satellite observations have demonstrated that all stars continuously expel matter, in varying quantities.

The phenomenon is more striking in stars of large mass, where it can be seen that losses of one hundredth-thousandth of mass occur each year. Interplanetary probes such as *Ulysses* and *SOHO* have measured a particle density (notably protons and electrons and traces of nuclei of helium and heavier elements) of between ten and one hundred units per cubic centimeter, with velocities of between 200 and 900 km/s (about 100 to 600 miles/s). These particles "carry" the coronal magnetic field with them. The rotation of the Sun forces the individual particles to describe curved trajectories, which is why the magnetic field's force lines are also arranged along curved lines in a spiral shape.

The greater the distance from the Sun, the lower the density of the solar wind. The heliosphere, the region in which the solar wind is still discernible, ends where the pressure exerted by the solar wind is equal to the pressure of interstellar gas, which is also a carrier of magnetic fields.

It has been calculated that the solar wind can be detected at a distance from the Sun equal to 100 times the distance between the Sun and the Earth (expressed as one astronomical unit, which is the equivalent of approximately 150 million kilometers or 93 million miles) and therefore well beyond the orbit of Pluto.

Satellite observations have made it possible to observe the most intense line emitted by hydrogen, known as Lyα or Ly-a, which occurs in extreme ultraviolet. Since this is the most intense emission from the uni-

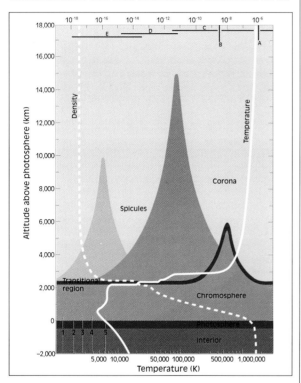

Left: This diagram shows the variation in temperature (solid line) and in density (broken line) in the solar atmosphere. The vertical scale on the left shows the height above the photosphere in kilometers. The horizontal scale, below, shows temperature; the horizontal scale at the top, density. Prominences and filaments extend through the chromosphere and the corona. Everything is immersed in the solar wind.

verse's most plentiful atom, it has also been possible to measure the density and movements of a gas as rarefied as interstellar gas. This led to the discovery of the existence of an interstellar wind that flows inside our Solar System, extending to within about a hundred astronomical units from the Sun. Although it has a density of one atom of hydrogen per ten cubic centimeters of space, the interstellar wind interacts with the solar wind, elongating the heliosphere in the direction of its flow.

The solar wind and Earth

Interplanetary probes have enabled astrophysicists to study the solar wind in the vicinity of the Earth. It is subject to considerable variation,

SOLAR OBSERVATION

The Sun was one of the first objectives of space astronomy. From the end of the 1950s onwards various families of satellites and probes were launched by the United States and the Soviet Union to investigate how the mechanisms of the star actually work. After NASA's Pioneer and OSO probes and Soviet Russia's Cosmos and Prognoz series of probes, the 1970s saw two large solar observatories come into service: one was installed on Skylab, NASA's first orbital laboratory, and the other on the Russian space laboratory, Salyut. While these great orbital laboratories were fulfilling their tasks, development continued on solar probes and satellites which were larger and more advanced compared with their predecessors. At the beginning of the 1980s, attention centered on the Ulysses probe's mission, built by the ESA (European Space Agency) with the aim of carrying out close investigation of the Sun's polar regions for the first time.

On October 6, 1990, Ulysses was inserted into an orbit around the Earth; it then used its own powerplant to re-launch itself towards Jupiter, approaching this planet two years later. The Ulysses probe's trajectory used a gravity assist maneuver, exploiting the planet's gravitational force, and changed course, heading towards the Sun, observing

Below: A drawing of the Ulysses probe and its engines immediately after its release from the bay of NASA's Discovery shuttle which had taken it into orbit in October 1990. Ulysses weighed 370 kg (815 lb) and had a payload space of 3 x 3 meters (10 x 10 feet). It reached Jupiter in February 1992 and then flew close to the poles of the Sun in 1994 and 1995.

the solar south pole in June 1994 and the north pole the following year. The mission was extended and in 2000–2001 Ulysses again flew over the Sun's polar regions. In November 1986, ESA and NASA signed an important international co-operation agreement, known as the Solar-Terrestrial Science Program (STSP), in order to study the origin and transmission of the solar wind, its interaction with the Earth's magnetic field and the resulting effects on the terrestrial atmosphere over a period of time.

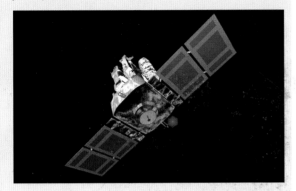

This project involved the launch of the SOHO (Solar and Heliospheric Laboratory) probe, which temporarily escaped from the control of the ground station during a normal in-orbit maintenance operation. SOHO was at one point given up for lost but after an amazing search and rescue operation, it was gradually brought back under control and the solar observatory resumed its interrupted mission. At the time of its "disappearance" it had already acquired 2 million images of the Sun and had reached the end of its originally estimated operational life.

Below: The Ulysses probe during the land-based tests carried out at the ESTEC-ESA Nøordwijk Center in the Netherlands. The entire space vehicle was built by a European industrial consortium led by the German company, Dornier, with an Italian company, Laben, being responsible for construction of the probe's central processor.

Left: A drawing of the SOHO probe, launched in December 1995 by an Atlas Centaur vector rocket. SOHO became operational in March 1996, and weighed approximately 2 tons; with its solar panels deployed its span was nearly 8 meters (26 feet).

Below, left: The Japanese satellite Hinotori built by NEC Corporation for ISAS which launched it into orbit from the Kagoshima Space Center in February 1981 using a Japanese M-3C-2 vector rocket. The Hinotori satellite weighed 180 kg (400 lbs) and was inserted into a near-circular orbit (radius varying from 580–640 km/360–400 miles).

even within the course of a few days and this is dependent on the Sun's rotation. Depending on the feature of the solar surface that happens to be facing towards Earth, the magnetic features, and the energy of the particles emitted, also change. When the Earth is exposed towards a solar area on which there is a coronal hole, a region in which the force lines of the magnetic field are not closed but directed radially into interplanetary space, streamers of very fast particles reach our planet. Other, occasional events of brief duration, such as flares, give rise to emissions of energy particles which, when they reach Earth, disturb its ionosphere.

Radio emissions
Various forms of solar activity are typical of the different levels of the Sun's atmosphere. Sunspots, granulation and faculae can be seen at pho-

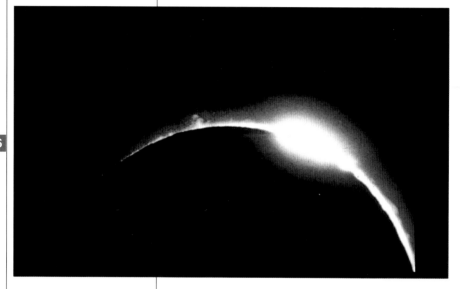

Above: The solar chromosphere observed during a total eclipse. Three small prominences can be seen as well as a larger one along the pale pink arc of the chromosphere.

tospheric level; prominences and flares at chromospheric level; variable magnetic structure, with hotter regions and coronal holes, at the level of the corona. All these phenomena are, generally speaking, interconnected and the solar cycle reveals that a maximum of sunspots is accompanied by an increase in the number of prominences and more frequent flares. One aspect of coronal activity: radio emissions, clearly demonstrates the existence of these correlations.

The Sun is, in fact, a radio source: this fact was discovered in 1946, during a very busy period of maximum solar activity, due in part to progress that had been made in the field of radio transmissions resulting from the military demands of World War II. Ionized gas, of which the chromosphere, the transition zone and the corona are composed, is the source of solar radio emissions. These radio emissions can be described in terms of "noise"; they can be equated with the rustles, crackles and whistles that are heard when the volume of a radio is turned up too high or when there is interference with transmissions by meteorological or magnetic storms. The noise is at a more or less constant level when the Sun is in one of its quiet periods. During periods of solar activity, however, the background noise is overlaid by other noises of a different nature, also associated with phenomena that take place at the levels of the photospheric and the chromospheric layers.

Birth, life and death of the Sun

Like all stars, the Sun has a life and must therefore also have an end. The Sun was formed approximately 5 billion years ago as a result of the contraction of a cloud of gas and interstellar dust. Gradually, as the nebula

Left: In his book on the Sun, Secchi depicts solar prominences observed in the Hα line of hydrogen, showing their many various shapes.

Above: An illustration by the Jesuit priest Christoph Scheiner plotting the track taken by two sunspots across the solar disk in June 1626 and January 1627. The distinction between the umbra and the penumbra, and the change in the sunpots' apparent shape as they near the limb of the Sun, demonstrated the influence of features of the Sun's surface rather than that of mysterious external bodies. The latter was a hypothesis supported by those who defended the Aristotelian idea of the heavens as incorruptible.

SUNSPOTS

In 1610, while Galileo was observing the Sun through his telescope, he discovered that a number of spots, varying in complexity and fairly persistent, although not permanent, were visible on its surface. The fact that these sunspots were not always present, and were changeable size and shape, provided Galileo with a very powerful and direct weapon in his polemic against the Aristotelian doctrine of physics which maintained that the heavens were "incorruptible."

The sunspots were still numerous and occurring frequently around 1626, but by 1640 there were fewer of them and they subsequently became a real rarity for about fifty years. They reappeared, in some numbers, at the beginning of the eighteenth century and from then onwards the phenomenon followed a cyclical progression, alternating between periods of maximum and minimum activity approximately every 11 years; the existence of a sunspot cycle was demonstrated by Schwabe, based on observations he had painstakingly collected for 25 years, between 1826–1851. The cycle was more fully documented by Wolf in 1852.

The variation in solar activity indicated by the sunspots came to be recognized as a cause of particular climatic phenomena on Earth. During

Left: The observation of sunspots using the indirect, projection method. By using this method, the Sun could be observed even in full daylight, whereas for direct observation with a telescope it was necessary to wait until the star was setting on the horizon. This drawing is taken from Scheiner's book Rosa Ursina, which was published in 1630.

Below: The Sun-Earth-Moon triangle, as depicted in a sixteenth-century book. The Sun is shown at a finite distance, and the Moon is shown in its first quarter when its angular distance from the Sun is less than 90°.

the years when there were virtually no sunspots, for example, Europe experienced a period of intense cold, known as the Little Ice Age of the second half of the seventeenth century. During this time glaciers advanced over the entire Alpine region, covering pastures, cutting communication routes and depositing moraines which today lie far below the terminal ice tongues of present-day glaciers. Were a connection between these two phenomena, thought probable by many scientists, to be proved, it would help to explain the origin of the great ice ages.

In 1769 Alexander Wilson turned his attention to an effect of perspective, connected with the sunspots' approach towards the limb of the Sun, which showed the area of their umbra to be at a lower level than that of the surrounding photosphere. As Scheiner had already suspected, the period of revolution of sunspots changes considerably according to their latitude: this was the first evidence of the fluid nature of the Sun, or at least of its outermost layer.

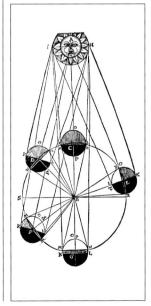

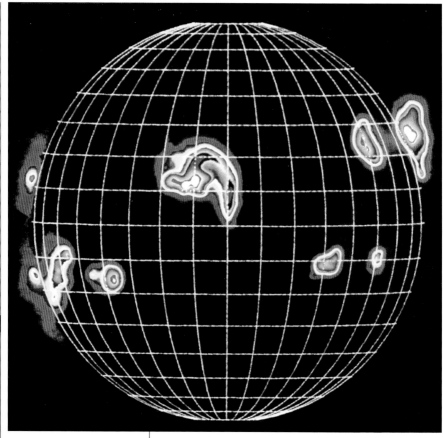

Above: Unlike the preceding pictures of the corona obtained with soft X-ray telescopes (i.e. comparatively low-energy X-rays), this picture, although also taken from Skylab, was obtained by telescopes capable of detecting hard X-rays (i.e. higher energy X-rays, corresponding to temperature values equal to, or in excess of, 5 million degrees). Only limited coronal regions emit hard X-rays; in order, therefore, to locate their position on the surface of the Sun, a grid has been inserted showing the solar parallels and meridians.

contracted under the action of its own gravity, it became more dense and less transparent. It became opaque, and was capable of retaining the heat produced during the contraction process. As a result, its temperature started to rise until it reached several million degrees in the center. At this point, the necessary conditions existed for the first nuclear reactions to be triggered. However, the pressure exerted by the gas particles was not yet sufficient to resist the force of gravity: this happened at the moment when the core temperature reached its present 15 million degrees Kelvin, when the Sun was approximately a hundred million years old. The star then stopped contracting and its mature phase had begun. Meanwhile, a disk of solid and relatively cold material had formed around the Sun from which, scientists have deduced, the planets were formed. Similar disks or rings have, in fact, been observed around stars that are younger versions of the Sun by means of the *IRAS* satellite: since these disks are formed of matter with a temperature of barely a few hundred

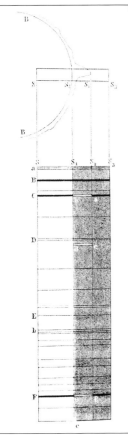

degrees Kelvin, these stars irradiate in the extreme infrared, and are inaccessible to terrestrial observation.

During the present phase of its maturity, the Sun is consuming its nuclear fuel of hydrogen, transforming it into helium. It has been calculated that in approximately another 5 billion years there will be no more hydrogen left at the center of the Sun, only helium: that is when the old age of the Sun will begin. At a temperature of 15 million degrees Kelvin, helium is inert. The supply of nuclear energy will have been exhausted. Without its supply of nuclear energy, the central core will cool and the pressure that counteracted the force of gravity will diminish: the Sun will start to contract, heating as it does so. A zone that is still rich in hydrogen surrounding the core will start to cause nuclear reactions; subsequently, when the core temperature reaches 100 million degrees Kelvin, the helium itself will start to change into carbon. Since the production of energy becomes far greater at such high temperatures than in the preceding phase, the Sun will have to increase its area of dissipation in order to irradiate all the energy produced. It will therefore undergo a process of expansion that will increase its radius to a value a hundred times greater than at present. Following this expansion the surface temperature will fall from the present 6,000 K to less than 3,000 K: the Sun will have become a giant red star and its photosphere will stretch out far enough for it to encroach upon the orbit

Left: By placing the slit of a sufficiently high-resolution spectroscope radially on the image of the Sun, it is possible to observe, without an eclipse, the inversion of the main spectral lines up to highest altitudes of the chromosphere and the prominences. The line indicated by the letter "d" was thought to be an element peculiar to the Sun, called helium, lacking in laboratory-created spectra: only later was it also found to be present on Earth.

Below: When the Moon completely hides the Sun during a total eclipse, the spectrum of the solar light around its limb is reversed: the dark lines become light. This phenomenon, discovered during the second half of the nineteenth century, is shown here in a color photograph taken in 1970. In the green can be seen a ring caused by the coronal iron emission line, ionized 14 times, at a wave length of 5303 Ångstroms and, in yellow, to the left of the D1 and D2 lines of sodium, the more intense D3 line of helium.

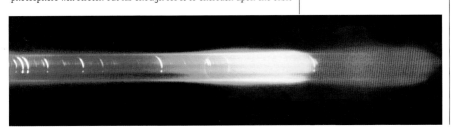

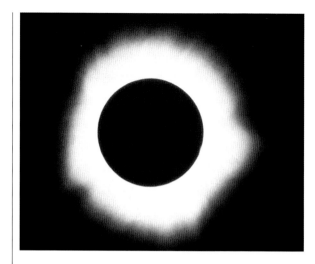

Right: A photograph of the solar corona observed in Australia during the total eclipse of 1921. This corona is typical of periods of maximum activity.

Below: Wilhelm Herschel's interpretation of the effect of perspective which results in an apparent shift of the umbra of sunspots in relation to the penumbra: he believed that a solid underlying crust, feebly illuminated due to the intervening gassy layers, could be glimpsed through holes in the fiery atmosphere that surrounds the Sun.

of Venus and perhaps that of Earth as well. Once this phase is over, the Sun will no longer be capable of releasing any more energy. The rarefied external layers surrounding it will slowly dissipate into interstellar space and the hot and weak central core will slowly cool. In this way the Sun will have reached the final stage of its life; it will have become a white dwarf star, small in size but very hot. Tens of billions of years will pass before the Sun will cool completely, changing into a black dwarf star, a "Sun" that will become invisible.

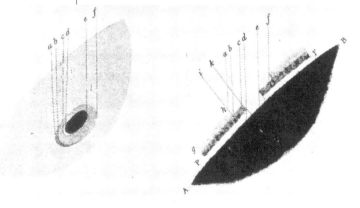

THE STUDY OF SOLAR PHYSICS

Until the second half of the nineteenth century, ideas about the physical composition of the Sun were very vague. It is reasonable to say that the physical study of the Sun started with two, almost contemporary, measurements of the heat irradiated by it: 3.86×10^{33} ergs per second, equal to 3.86×10^{23} kW; these values were obtained in 1827 by the French physicist, C.M. Pouillet and by John Herschel. They were indications of huge quantities of energy, far greater than any that could be produced by the most violent chemical reactions and it was therefore necessary to hypothesize the existence of an energy source of hitherto unimagined power.

In 1848, the German physicist Julius Robert Mayer advanced the hypothesis of a continual rain of meteorites, but the increase in the Sun's mass that this would have implied would have caused a small but perceptible reduction in the length of the solar year. In 1854 Hermann von Helmholtz suggested an answer when he calculated the

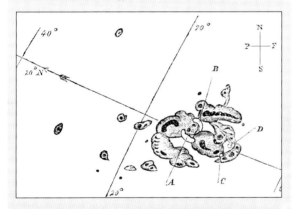

Above: The solar corona, photographed during the total eclipse of December 12, 1871. Bearing in mind that this was taken only 45 years after photography had been invented, it is a very good picture.

Left: In 1859, Carrington discovered flares in white light, flashes of light that suddenly flare up against the dark background of sunspots, indicated here by the letters A, B, C and D. They were originally thought to be caused by heat that was released following the fall of fragments of large meteorites on the Sun. In the mid-nineteenth century, the hypothesis of a "rain" of meteorites was still being put forward as an explanation, albeit partial, for the star's constant supply of energy.

amount of heat released during the contraction and increase in density of the solar material from a less dense state: this produced a sufficient quantity of heat to warm up the Sun's mass to a temperature of tens of millions of degrees and to store up the quantity of heat necessary to keep it supplied with energy for 20 million years.

In the mid-nineteenth century, however, utter confusion reigned as to the value of the temperature of the surface strata; this lasted until 1878 when Joseph Stefan calculated the correct relationship between a body and the amount of heat it emits; he came close to the correct value of approximately 6,000 degrees absolute (K).

THE INNER PLANETS

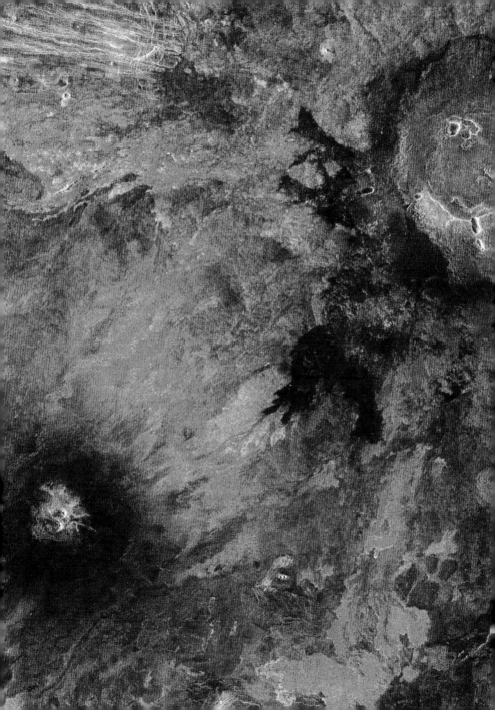

MERCURY

Below: An image of the Mercury's northern hemisphere taken by Mariner 10 at a distance of 55,000 km (34,000 miles) from the planet's surface.

Physical characteristics

Mercury is the planet closest to the Sun and the smallest of the terrestrial planets, the others being Venus, the Earth and Mars. Its radius and mass are, respectively, 0.382 and 0.0553 times the Earth's radius and mass, and its mean density is almost equal to that of the Earth, being equivalent to 5.43 times the density of water. Mercury's orbit is very elliptical and is inclined by approximately 7° to the plane of the ecliptic. These are two features only shared by Pluto: all the other planets move along virtually circular orbits, at only a slight angle to the plane of the ecliptic. Where it not for these two exceptions at opposite boundaries of the Solar System, this could have been described as lying on virtually a single plane. Mercury's mean distance from the Sun is 2.58 times less that the distance separating the Earth from the Sun and a quantity of solar radiation 6.67 times greater reaches Mercury's surface than the terrestrial surface. Mercury completes an entire revolution around the Sun in 87.967 days.

For a long time it was thought that the period of rotation was equal to that of its revolution. If this had been the case, Mercury would have the same hemisphere permanently turned towards the Sun, in the same way as the Moon always faces the same way towards the Earth. Only as recently as 1965 did radar observations determine Mercury's axial rotation period as 58.65 days, exactly two-thirds of its period of revolution, a phenomenon which celestial mechanics explains as a consequence of its high orbital eccentricity. Were Mercury's orbit circular in shape, the rotation period might indeed have been equal to the period of revolution.

Atmosphere

Unlike the Earth, Venus and Mars but similar to the Moon, Mercury has virtually no atmosphere. A planet's ability to retain the volatile, gassy substances which would constitute its atmosphere depends on its mass and on the surface temperature. In fact, the higher the temperature, the greater the speed with which the gas particles move around, and if their speed is near or greater than the escape velocity of the planet, the atmosphere escapes into interplanetary space. The greater the mass and the smaller the radius, the greater is the escape velocity. Mercury's small mass and the high temperature have resulted in mere traces of an atmosphere being retained by this planet, consisting of minute quantities of hydrogen, helium, oxygen, sodium, potassium and argon. The mean temperature and pressure values at ground level are, respectively about 440°C (820°F) and a millionth of a billionth of atmosphere. The latter value gives some idea of the almost total absence of atmosphere on Mercury. On Venus, Earth and Mars the atmosphere functions as a temperature regulator; Mercury's lack of atmosphere means that on the illuminated surface, the temperature rises above about 430°C (800°F), while on the dark side it falls rapidly to 185°C below zero (–300°F).

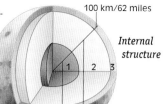

Internal structure

100 km/62 miles
1,800 km/1,100 miles
1,500 km/ 900 miles

1—Core of iron and nickel
2—Rocky mantle
3—Thin crust

Atmosphere

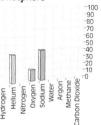

Hydrogen, Helium, Nitrogen, Oxygen, Sodium, Water, Argon, Methane, Carbon Dioxide

Characteristics of Mercury

Mean distance from Sun (AU)	0.39	Mass (g)	3.303 x 10²⁶
Mean distance from Sun (10⁶ km)	57.91	Mass (Earth = 1)	0.055
Orbital period (days)	87.969	Equatorial radius (km)	2,439
Mean orbital velocity (km/s)	47.89	Equatorial radius (Earth = 1)	0.382
Orbital eccentricity	0.2056	Mean density (g/cm³)	5.43
Apparent mean diameter of Sun	1°22'40"	Mean density (Earth = 1)	0.98
Inclination of orbit to ecliptic (°)	7.004	Volume (Earth = 1)	0.056
Number of satellites	0	Ellipticity*	0.0

Equatorial surface gravity (m/s²)	3.78
Equatorial surface gravity (Earth = 1)	0.39
Equatorial escape velocity (km/s)	4.3
Sidereal rotation period at equator	58 days 15 h 36 min
Inclination of equator to orbit (°)	2

*Ellipticity is (Re—Rp)/Re, where Re and Rp are the planet's equatorial and polar radii, respectively.

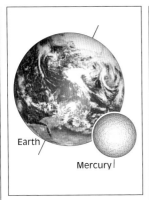

Above: A comparison of the size of Mercury and Earth and the inclination of their rotational axes.

Internal Structure

Due to the differences in temperature between the innermost and outermost regions of the primordial cloud, the internal planets were formed when zones rich in silicates and ferrous materials condensed, while in the outermost zone where the giant planets were formed, frozen water, methane and ammonia were also present.

The high mean density of Mercury, second only to that of Earth, is an indication of the existence of a large core of iron and nickel that extends for approximately seven-tenths of the planet's radius. Mercury is, therefore, the most iron-rich planet in the entire Solar System. It is also the only terrestrial planet, apart from Earth, to possess a bipolar magnetic field, its strength equal to approximately one-sixth of Earth's magnetic field; as in the case of the Earth, the axis of the magnetic field is inclined by approximately 11° to the rotational axis.

The presence of a magnetic field presupposes the existence of an electrically-conductive molten core: from this it has been deduced that the interior of Mercury contains a fluid zone surrounding the central core, but the thickness of this is as yet unknown.

Above, left to right: Images of Mercury taken by Mariner 10 from distances of, respectively, 3.5 million km (2.17 million miles) and 1.84 million km (1.14 million miles), 952,000 km (591,000 miles) and 500,000 km (310,000 miles). The vast, luminescent area already distinguishable in the first image is 480 km (300 miles) in diameter.

The enigma of Mercury's polar regions

A group of researchers at the Jet Propulsion Laboratory in Pasadena, California discovered through radar observation that radio waves transmitted from Earth are reflected by areas around Mercury's poles with an intensity that would suggest that these regions have deposits of ice. Mercury has no atmosphere, however, and therefore the presence of water would seem to be inexplicable; during the first billion years of the planet's life, moreover, the high temperatures caused by its nearness to the Sun would in any event have soon caused any water present on the surface to change into water vapor which would have escaped into interplanetary space, since Mercury's weak gravitational attraction would not have been sufficient to hold onto it.

THE BRIGHT PATCHES AND DARK AREAS OF MERCURY

In 1881, Giovanni Schiaparelli was the first astronomer to claim that he had made out details on the surface of the planet, while studying Mercury from the Brera astronomical observatory in Milan, Italy. He was under the impression that these details always remained in the same position in relation to the terminator (the zone of separation between the dark and light parts of the disk). From this he concluded that the planet always has the same face turned towards the Sun, rotating on its own axis for the same amount of time and in same direction in which it revolves around the star. Schiaparelli also published a map of what he assumed was its sunlit hemisphere. These observations were confirmed by all subsequent astronomers, notably by Antoniadi, based on his observations carried out during the years 1924–1929, and by Bernard Lyot and Andouin Dollfus who observed the planet during the years 1942–1944 from the Pyrenees. These conclusions concerning the rotation period were, however, inaccurate and the details purportedly observed were, at least in part, a product of the imagination.

In 1965, the first radar observations made with the Arecibo telescope in Puerto Rico showed that the planet rotates on its axis with a period equal to two-thirds of the revolution period: that is to say, 58 days. This means that the illuminated hemisphere is not always the same. In defense of the visual observers, it must be said that under the best conditions of visibility, when the planet is furthest away from the Sun, Mercury does have the same hemisphere turned towards Earth, and an observer's attention is inevitably drawn to the most obvious details which appear in the form of bright patches and dark areas.

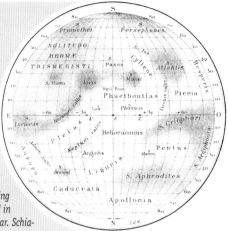

Above: Until 1965 it was believed that the same hemisphere of Mercury, illuminated by sunlight, was always visible from Earth. This map was drawn by Eugenios Antoniadi in 1929 to show the planet's features. When the first observations made by radio-astronomy revealed that the planet does not always have the same face turned towards the Sun, his nomenclature was dropped.

Below: A drawing by Antoniadi, showing the comparative sizes of Mercury (left) Mars (center) and the Moon (right). This drawing also underlines the similarity between the general appearance of the Moon and that of Mercury, subsequently confirmed during space missions.

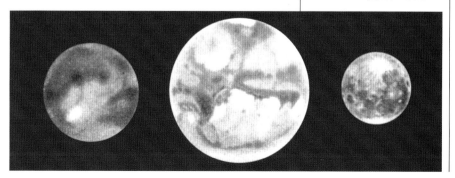

Measurements carried out by radar indicate that around Mercury's North Pole there are more areas with ice-like deposits than at its South Pole.

Above: A drawing of the Mariner 10 probe, launched on November 3, 1973 by an Atlas Centaur vector rocket. The probe reached Mercury on March 29, 1974.

Temperature

The absence of atmosphere means that temperature varies greatly from the equator where it reaches 430°C (about 800°F), and the poles, where the mean value is 135°C (210°F) below zero. Although low, these temperatures are not sufficiently low to allow of the presence of ice, which would vaporize. Temperatures of 160°C (about 260°F) below zero are needed if evaporation is to be limited to only one meter (3.28 feet) every billion years. Perhaps such low temperatures may exist deep within the large impact craters, since their walls would block out sunlight. The water source could still be contained within the interior of the planet: if this were to be confirmed by future observations, it would be one of the most surprising and unexpected discoveries in space exploration.

MARINER 10'S "SAIL-POWER"

Plans for the exploration of Mercury were both conceived and developed shortly after the first men landed on the Moon. It was a difficult period for NASA, and for its funding, because public interest in the exploration of space had begun to wane.

It was during this time that studies were being undertaken at the Jet Propulsion Laboratory in Pasadena, California, into a type of orbital technique known as gravity assist which would exploit Venus's gravitational action to send a projected probe towards Mercury. The mission was given the go-ahead in 1969 and work started on building Mariner 10, *the sixth and most advanced space vehicle in the* Mariner *series of probes which incorporated an extraordinarily high number of innovations compared with its predecessors. These included a more powerful onboard computer which could be re-programmed from Earth; a shield in the shape of a parasol made of glass fiber and teflon to protect the vehicle from the intense solar radiation, as well as a rotating mechanism which adjusted the position of the two solar arrays to control heat absorption. On November 3, 1973 the probe was launched on its journey towards Venus, which it reached the following year after a voyage full of crises, all of which were fortunately resolved. Once the gravity assist maneuver had been successfully achieved,* Mariner 10 *was set on its new course towards Mercury. After a major malfunction, the solar panels had to be repositioned so that they could act as sails which, under pressure from solar radiation, were to provide thrust and direction for the probe's flight. On March 29, 1974 the probe made its first flight over Mercury's surface. The images sent back to Earth revealed striking similarities with the Moon's environment. During the subsequent flybys* Mariner 10 *measured the planet's weak magnetic field and sent more images of the surface (some with a resolution of only 100 meters/328 feet) back to Earth.*

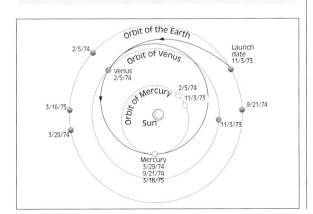

Top: An image of Mercury taken by Mariner 10 *from a distance of 20,700 km (12,850 miles). It shows a relatively young impact crater in the center of an older crater.*

Above: *This crater, observed from a distance of 31,000 km (19,250 miles) is 100 km (62 miles) in diameter. Radial striations can be seen, probably produced by material ejected from the crater (upper left).*

Left: *The orbital pathways taken by* Mariner 10.

VENUS

Above: A computer-generated mosaic of a Magellan *radar image of the northern hemisphere of Venus, produced at the Jet Propulsion Laboratory in Pasadena, California. The color is based on color photographs of the Venusian surface taken by the Venera 13 and 14 probes in 1982.*

Physical characteristics

Venus is the planet whose orbit brings it closest to Earth and it appears to us as the brightest star in the sky at dawn and dusk. Its luminosity is not only due to its nearness to Earth, but to its *albedo* or capacity to reflect sunlight. The albedo of Venus is the highest in our Solar System at 0.76 which means that the planet reflects 76% of the sunlight that it receives (incident sunlight). In comparison, Earth has an albedo of 0.39, while the albedo of Mercury is only 0.06 and that of Mars 0.15. The albedo is increased by the presence of clouds, which are highly reflective, and diminished by the total or almost-total absence of atmosphere which has the opposite effect. The mass, radius and mean density of Venus are almost equal to terrestrial values.

Venus's surface rotation period is 243 days and the direction of rotation is from east to west. This is a peculiarity which Venus shares only with Uranus: all the other planets rotate from west to east, in the same direction, that is, as they move around the Sun. This predominant direction of rotational movement in the Solar System has come to be known as "direct" or "prograde" motion and its opposite "retrograde" motion.

Venus's rotational period is the time that the planet takes to accomplish a complete turn on itself, about its axis, measured in relation to a fixed point of reference, such as the stars. The speed of rotation is a factor in determining physical conditions existing in its interior, such as the presence and intensity of its magnetic field. The alternation of day and night on Venus, however, depends on the time that the planet takes to complete an entire revolution around the Sun. In the case of Venus, and also of Mercury, the period of rotation and of revolution are comparable, unlike all the other planets which have a rotational period that is much shorter than their orbital period. In consequence, the solar day on Venus is equivalent to 116 terrestrial days (and on Mercury, to 176 terrestrial days). Another distinguishing feature of Venus is that it describes an almost perfectly circular orbit.

The belief that a planet once thought to be so similar to Earth could support some form of life has been shown to be unfounded by the observations carried out by Soviet and U.S. space probes, which revealed an extremely hostile environment on Venus.

The interior of Venus is also likely to differ substantially from that of the Earth, despite the fact that these two planets have very similar mean densities, Venus's being 5.25 g/cm3 and Earth's 5.52 g/cm3. In fact, unlike the Earth, Venus does not possess a magnetic field. It is possible that were the Earth to turn as slowly as Venus, it too would have a magnetic field too weak to be measured.

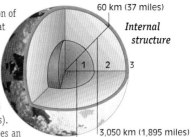

Internal structure

1—Core of partially molten metal
2—Mantle
3—Crust

Atmosphere

The chemical composition of the Venusian atmosphere differs totally from that of the Earth. Carbon dioxide accounts for 96.5%, nitrogen for 3.5%,

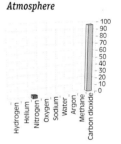

Characteristics of Venus

Mean distance from Sun (AU)	0.72	Mass (g)		4.870 x 10^{27}
Mean distance from Sun (10^6km)	108.20	Mass (Earth = 1)		0.815
Orbital period (days)	224.701	Equatorial radius (km)		6.051
Mean orbital velocity (km/s)	35.03	Equatorial radius (Earth = 1)		0.949
Orbital eccentricity	0.0068	Mean density (g/gm³)		5.25
Apparent mean diameter of Sun	44'15"	Mean density (Earth = 1)		0.95
Inclination of orbit to ecliptic (°)	3.394	Volume (Earth = 1)		0.857
Number of satellites	0	Ellipticity*		0.0

Equatorial surface gravity (m/s²)	8.60
Equatorial surface gravity (Earth = 1)	0.88
Equatorial escape velocity (km/s)	11.18
Sidereal rotation period at equator	243 days 0 h 14.4 min
Inclination of equator to orbit (°)	177.3

*Ellipticity is (Re—Rp)/Re, where Re and Rp are the planet's equatorial and polar radii, respectively.

Below: These three drawings provide a schematic illustration of the phenomena that occur in the Venusian atmosphere due to rotation and differences in temperature at various latitudes. "Hadley cells" denote the regions defined by gas circulation that rises in hot regions and flows into cold regions; once the gas has cooled, it falls into the hot regions again (first drawing on left). Venus rotates slowly with a retrograde motion. The upper atmosphere, however, rotates far more quickly, completing a turn in 4 days. As a result of this, the winds blow from east to west at all latitudes, but their speeds increase from 1 meter (3.28 feet) per second at ground level to 100 meters (328 feet) per second in the highest cloud layers (second drawing). Corresponding to the pole are two clear patches rotating around it (third drawing) possibly caused by planetary step waves, of unknown origin.

and traces of sulfur dioxide, argon, carbon monoxide and oxygen are present: a highly toxic atmosphere for any terrestrial life forms.

A good illustration of the density of the Venusian atmosphere is provided by this comparison: Venus's atmosphere represents one ten-thousandth of the planet's total mass, while the terrestrial and the Martian atmospheres measure one millionth and one ten millionth of their respective planets' masses.

The greenhouse effect

It is thought that originally Venus's surface was, like that of the Earth, also partially covered by oceans and seas. The Sun, which was a young star at that time, gave out a lesser quantity of energy than it does now and the amount of heat absorbed by the surface of Venus equaled the amount that the planet re-emitted. Gradually, however, as the Sun slowly reached its phase of "maturity" the quantity of solar radiation emitted increased and as a result the surface of the planet began to heat up more quickly; its water started to evaporate and carbon dioxide began to be released from the ground, helping to increase the opacity of the atmosphere and creating a strong greenhouse effect.

The expression "greenhouse effect" is an apt one and descriptive of what is actually taking place. The glass roof of a greenhouse allows the entry of visible solar radiation which corresponds mainly to yellow-green

Right: Two views of the Venusian sphere. The uppermost layer of the planet's cloud blanket was photographed at seven-hour intervals. The arrows point to the same region on both images. From the movements it has been possible to deduce that the rotation period of the upper atmosphere is 4 days.

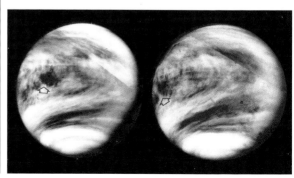

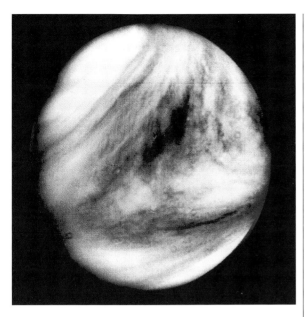

Left: An image of the thick atmosphere surrounding Venus, taken by the Mariner 10 probe: this was taken from a distance of approximately 2.75 million kilometers (1.70 million miles) and shows a dark, y-shaped band near the equator.

Below: A comparison of the size of Earth with that of Venus and the inclination of their rotational axes.

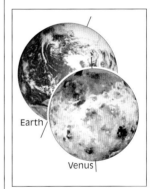

bands of the spectrum, characteristic of a body having a surface temperature of approximately 6,000 K, and this is indeed the surface temperature of the Sun. It warms the ground and the plants inside which, in turn, radiate heat. However, since their mean temperature is 300 K or slightly higher (approximately 27°C/80.6°F) they re-emit in the infrared band, to which glass is opaque. The solar heat is therefore trapped within the closed space of the greenhouse and the temperature rises.

A greenhouse effect originated in the same way on Venus when solar heat was still able to penetrate the thick layer of carbon dioxide of the planet's atmosphere, heating its surface, but when the atmosphere subsequently became opaque to the infrared radiation re-emitted by the ground, its temperature rose rapidly, and all the planet's water underwent a process of evaporation. It is calculated that this could have happened over a period of time lasting approximately 300 million years, relatively short when compared with the age of the Solar System at 4.5 billion years. Something similar, albeit in a far less drastic fashion, could happen on Earth, if the carbon dioxide content in our atmosphere and the destruction of the Earth's great forests and jungles continues to increase. Trees and plants are our natural protection against an increase in the greenhouse effect because they have the capacity to absorb carbon dioxide during the process of photosynthesis.

Temperature

The density of this planet's atmosphere is also responsible for virtually negligible variations between the Venusian daytime and night-time temperatures; even the variations in illumination by day and by night are

Above: Two globes showing the topography of Venus in false colors. The mean level (which on Earth is represented by sea level) is shown in bright blue. The depressions are shown in dark blue; green, yellow and red denote land of increasing height. The globe on the left shows the south pole and the "continent" of Aphrodite, straddling the equator. The globe on the right shows the north pole and Ishtar Sierra, near the top. The two circles containing no detail cover polar regions which Pioneer-Venus did not observe.

very small and this means that Venus only experiences darkness or semi-darkness.

The variations of temperature between the equator and the poles are also only a matter of a few degrees. However, above the first layer of cloud, at an altitude of approximately 100 km (62 miles) from the ground, the differences in temperature between day and night become very noticeable, equivalent to approximately 300 K on the illuminated side (or 27°C/80.6°F above zero) and 130 K on the dark side (or 143°C below zero/–225°F). Other surprising features of Venus's atmosphere include the rapid rotation of the clouds, very strong winds at altitudes high above the surface and violent electrical discharges. The upper region of Venus's atmosphere is formed by three cloud layers that are very distinct from one another: the lowest layer is approximately 1 km (just over ½-mile) thick, the intermediate layer occurs between altitudes of 50 and 57 km (31 and 35 miles), and the top layer is more tenuous and is found between altitudes of 57 and 66 km (35 and 40 miles). Above and beyond these rises a thin, hazy layer that may possibly extend up to an altitude of 90 km (56 miles).

The winds at the level of the clouds blow from east towards the west in the same direction as the planet rotates and they tend to shift slightly with a spiral movement towards the poles where they form a vortex. Under the cloud layer the winds lessen in strength, with a speed at ground level of only about 1 m/s (3.28 feet/s).

The surface

All that we know about the surface of Venus has been revealed from analysis of radar signals transmitted by interplanetary probes and smaller probes that descended through the dense atmosphere until they touched down on the planet's searingly hot surface.

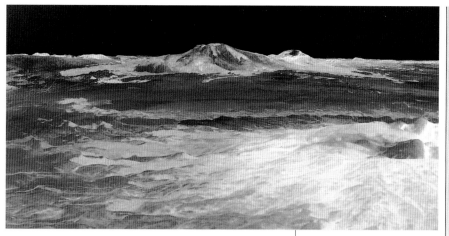

To date it has been possible to map 97% of the planet's surface, resolving features down to approximately a hundred meters (about three hundred feet). On Venus, as on Earth, there are highlands and depressions in the terrain; but a feature of Venus is that approximately 65% of its sur-

Above: A three-dimensional reconstruction of part of Atlas Regio. The Maat Mons volcano and Ozza Montes can be seen in the background, and the peak of the Sapas Mons volcano in the foreground. At the foot of the volcano and down its sides lava flows extend for hundreds of kilometers.

Left: The image and its related graph show the speeds of winds at various latitudes, measured in relation to the rotation of the planet in three different years.

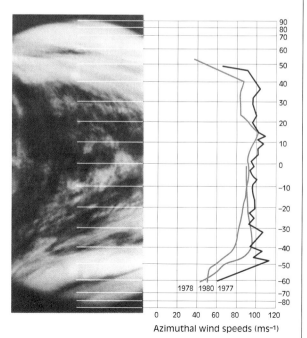

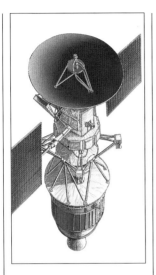

Above: The Magellan probe.

Right: A vertical section of Venera probes' landing module.
1. Atmospheric density gauge
2. Cloud aerosol extent and concentration gauge
3. Mass spectrometer
4. Gaseous phase chromatograph
5. Ultraviolet spectrometer and photometer
6. Telephotometer
7. Ground resistance gauge
8–9. X-ray fluorescent emission spectrometer
10. Soil sampling collection system
11. Temperature and pressure gauge
12. Radio-spectrometer
13. Moisture and pressure gauge
14. Solar batteries
15. Antennae
16. Containers
17. Aerodynamic panel
18. Insulation
19. Shell
20. Bracing structure
21. Base ring for landing

THE MOST WIDELY-EXPLORED PLANET

Venus holds the record for the number of planetary explorations. After a series of failed attempts by the United States and the Soviet Union, the first probe to make a successful approach to the planet, at a distance of only 34,830 km (21,643 miles) was the US Mariner 2 *in 1962. In October 1967 two reconnaissance missions to Venus took place almost simultaneously: launched by the United States, the* Mariner 5 *probe flew over the planet, transmitting data when it was only 3,990 km (2,477 miles) away. The Soviet Union launched* Venera 4, *with a small module attached to a parachute that undertook the first analyses of the constituents of Venus's blanket of cloud cover. The information it gathered led to the launch of the* Venera 7 *probe in 1969 which released a module that actually landed on the surface and continued to transmit data for 23 minutes, recording a pressure of approximately 90 atmospheres and a temperature of 475°C (about 900°F). Only in 1973, however, were scientists successful in obtaining a clear picture of the planet through the images transmitted by the* Mariner 10 *probe's television cameras when it*

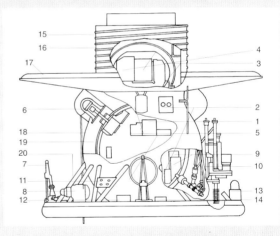

flew over the planet on its way to Mercury. Significant progress was made in 1975, through the second-generation of Venera *Russian space probes which sent the first photograph of the planet's surface back to Earth and mapped variations in temperature in Venus's northern hemisphere.*

The Americans carried out fewer missions than the Soviets but with significant results: NASA launched Pioneer-Venus 1 *in 1978 which mapped 93% of the surface of Venus and also detected volcanic activity, and* Pioneer-Venus 2 *which carried four conical mini-probes: these traveled through the planet's cloud layer at three different locations, analyzing the Venusian atmosphere before landing.*

The Magellan *probe was launched with the* Atlantis *shuttle from Cape Canaveral in 1989 with the aim of mapping 70–90% of the surface of the planet. In 1990 the* Magellan *probe entered a near-polar orbit around Venus. At the lowest point of its orbit the probe spent 37 minutes mapping a 25-kilometer (15-mile) wide swath of the planet's surface and then, on reaching the highest point of the orbit, it pointed towards the terrestrial listening stations and transmitted the data it had gathered: this was made possible by a unique parabolic radar antenna, 3.7 meters (12 feet) in diameter. For this radar mapping, continual adjustments had to be made to the probe's position. By January 5, 1992 the* Magellan *probe had sent back sufficient data to build up a map of 97% of Venus's surface. The probe also compiled a topographical map using radar alimetry with a resolution of approximately 30 meters (about 100 feet), and confirmed the gravitational field previously measured by the Pioneer-Venus probes.* Magellan's *activities made it possible for "new" Venusian landscapes to be mapped, and it gathered a greater quantity of data than the total collected by all six previous probes sent to Venus by NASA.*

Above: *The* Magellan *probe during ground testing before launching with the shuttle.*

Left: *The* Venera 15 *probe nearing completion of the assembly phase prior to its launch in June 1983.*

Below: *The plot of orbital pathways taken by the* Magellan *probe on its way towards Venus, finally entering a near-circular orbit around the planet.*

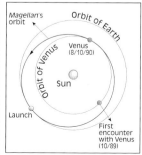

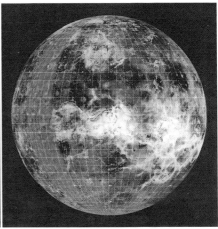

Above: These four hemispheres were obtained from a computer-generated photo-mosaic process using radar images taken by the Magellan probe, with additional data from previous Pioneer and Venus Orbiter missions to complete those parts that were not yet explored by Magellan. The superimposed grid represents a network of squares measuring 5 x 5 degrees and is centered (working from the hemisphere on the left) at 0, 90, 180 and 270 degrees longitude. These four images combined give a comprehensive view of the entire Venusian globe.

Right: Three large impact craters in the Lavinia region. The traces left by the expelled material can be clearly seen, as can their central peaks. The diameters of these three craters range from 37 to 50 km (23 to 32 miles).

face comes within the mean radius of the planet, 27% is 1 or 2 km ($^1/_2$ or $1^1/_4$ miles) below this, and only 8% rises as high as 11 km (nearly 7 miles) above it. In other words, the planet is covered with very extensive plains, with some depressions and few mountains, most of which are of very limited height.

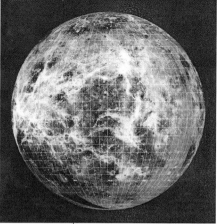

Two regions that rise above the mean elevation are Ishtar Terra in the northern hemisphere, almost as large in area as Australia, and Aphrodite Terra in the southern hemisphere, near the equator, with a surface area larger than that of Africa. In Ishtar Terra the Maxwell Montes reach an altitude of 11 km (nearly 7 miles). Maat Mons rises above Aphrodite Terra to an altitude of 8 km (just under 5 miles), and is particularly interesting because it is thought that this could be a volcano that is still active, as it is surrounded by relatively young lava.

There are a few basins which, were water still present on Venus, would be seas. The largest of these is Atalanta Planitia in the northern hemisphere. This forms a depression approximately 1.4 km (approximately $^3/_4$ mile) below the mean elevation of the planet and its area is equivalent to that of the Gulf of Mexico. Another interesting structure occurs in Terra Aphrodite: a canyon 2 km ($1^1/_4$ mile) deeper than the average planetary radius and 4 km ($2^1/_2$ miles) deeper than the adjacent ridges, with a maximum width of 280 km (174 miles); this has been given the name of the Diana Chasma (from the Greek, meaning abyss). Lastly, there is Sappho Mons, approximately 200 km (124 miles) in diameter, which exhibits structures that radiate outwards from it, suggestive of volcanic outflows. The variety in terrain and structures observed on Venus by the probes that have landed on the planet indicate that it has had a very complex geological history: volcanic craters and meteorite impact craters, as well as tectonic phenomena similar to those on Earth, have shaped the surface.

It is interesting to speculate what sort of landscape a hypothetical visitor to Venus would find (a prerequisite for the visit being the ability to survive its high temperatures and pressures). Apart from

Below: A detail of the surface of Venus obtained by the Magellan *probe in the Lakshmi region. Note the extensive network formed by two regularly aligned systems of tectonic features, almost perpendicular to one another.*

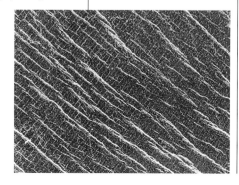

Above: The Sapas Mons volcano in Atlas Regio. This volcano is named after an ancient Phoenician goddess and is approximately 400 kilometers (250 miles) wide and 1.5 km (0.9 miles) high.

the face that perennial semi-darkness reigns, the very dense Venusian atmosphere would give rise to refraction phenomena that would be far more spectacular than the almost imperceptible ones produced by Earth's atmosphere.

For example, the different layers of varying density would produce a degree of refraction of solar rays that varies and it would be possible to see two or three images of the Sun, one for every layer. This is a phenomenon that can be observed on Earth as well, although only very rarely, when the Sun is very low on the horizon and a sudden variation in density occurs at a certain altitude in the atmosphere.

The interior of Venus

Analogies relating to mass, radius and mean density between Venus and the Earth demonstrate that these two planets have similar internal structures and that Venus, like Earth, possesses an iron-rich core.

From theoretical models it has been deduced that the thickness of the core and the mantle must be, respectively, 3,200 km (2,000 miles) and 2,800 km (1,700 miles), while the thickness of the crust would be approximately 20 km ($12^1/_2$ miles), less that the terrestrial continental crust, which is approximately 35 km (22 miles) thick, but more than the terrestrial oceanic crust which is approximately 10 km (6 miles) thick.

Magnetic fields

When we talk of planetary magnetic fields, we need to make a distinction between intrinsic magnetic fields and extrinsic magnetic fields. Intrinsic magnetic fields are generated by the movement of liquid conductive currents inside a celestial body which behave like gigantic natural dynamos: they are the norm rather than the exception in the larger bodies of the Solar System, from the Earth to the Sun. Among the smaller bodies, however, Mercury has a weak magnetic field, Venus does not have a measurable magnetic field and Mars possesses a weak magnetic field.

It is very important to find out the intensity of a planet's intrinsic magnetic field because from this can be inferred information about its interior, always the part about which we know least. It is, however, not

Above: Venus in conjunction with the Sun. The ring formed by the planet's atmosphere around the thin crescent is clearly visible, and can be seen both before and after the transit of the planet across the Sun's disk.

Below: A very clearly-defined impact crater on the planet's surface.

Above: A drawing from Il Saggiatore in which Galileo shows the phases of Venus; the planet's terminator was depicted with a jagged edge but this was an illusion caused by terrestrial atmospheric turbulence as suggested by the lunar terminator's similar appearance. This apparent unevenness made people jump to the conclusion that they were seeing "very high mountains."

Below: The phases of Venus in a sequence photographed in white light, taken by T.P. Pope and A.S. Murrel during the 1960s. The initial, thin crescent, parallel to the plane of the ecliptic, corresponds to the planet's conjunction with the Sun, when it has shifted slightly southwards.

easy to determine the exact intrinsic magnetic field since there is always an extrinsic magnetic field present as well, which is produced by the interaction between the solar wind, formed by ionized atoms and electrons, and the outermost and ionized atmosphere of a planet. The solar wind carries with it the Sun's own magnetic field. As in the case of the Earth, the uppermost regions of Venus's atmosphere, at an altitude of 140 kilometers (about 90 miles) and above, are ionized by solar ultraviolet radiation. This layer is called the ionosphere and it is a very efficient conductor of electricity. The surface where the pressure exerted by the solar wind equals that of the ionosphere is known as the ionopause: this occurs at an altitude of approximately 300 km (186 miles), and its effect is to deflect the solar wind around the planet, forming a barrier against penetration of solar particles into the atmosphere of Venus. The altitude

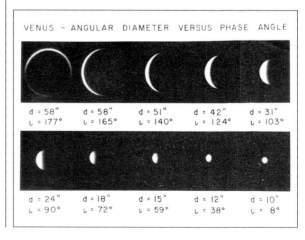

at which this barrier (or bow shock) occurs varies slightly with the solar cycle. The force lines of the magnetic field which extend behind the planet, form what is known as the "magnetotail."

Intrinsic magnetic field

Since the effect of the solar wind compresses the magnetosphere on the illuminated side of the ionosphere, increasing its intensity and drawing it out on the side that is not illuminated, the dark side is where it should be easier to detect any intrinsic magnetic field that does exist. Research in this area has been carried out by the US orbiting probe *Pioneer-Venus 2*. Based on the planet's mass and rotational speed, scientists expected to find a magnetic field equal to approximately three thousandths of the terrestrial magnetic field. However, observations indicate a much lower value, equal to one hundredth-thousandth of Earth's field.

Consequently, in contrast to the Earth, there is no evidence of a dynamo effect inside Venus. Nevertheless similarities in mass and density

Above: Another view of Venus, too distant to reveal its extremely hostile environment.

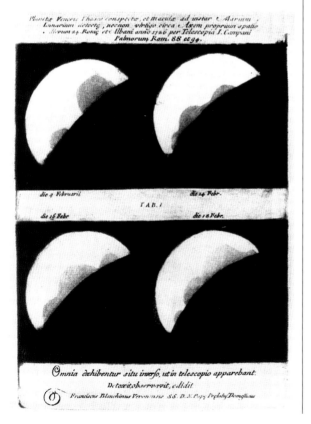

Left: Shading on the surface of Venus as observed by Francesco Bianchini in the early eighteenth century. The general appearance of these patches does not differ greatly from later photographic records but the conclusions which Bianchini, and others after him, drew from them as to the planet's rotation period were shown to be unfounded.

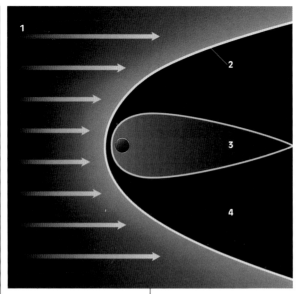

encourage scientists to infer that Venus must have a liquid core with conductive properties. It is possible that the mechanism that has set the Earth's dynamo in motion has not yet had the opportunity of triggering the same process in Venus, perhaps wholly or partly due to the planet's low speed of rotation.

All this means that the mysterious planet of Venus which is so similar to Earth in mass and size, can no longer hide aspects of its surface that remained secret and hidden for centuries despite mankind's curiosity, and its surface has now been explored by Soviet and US space probes which have revealed that it is a searingly hot wasteland.

Above: The solar wind hits the uppermost layers of Venus's atmosphere without being deflected since the planet has no magnetic field. The interaction of the solar particles with the ionized gas of the upper atmosphere produces a shock wave and an ionopause.
1. Solar wind
2. Shock front
3. Magnetopause
4. Magnetosphere

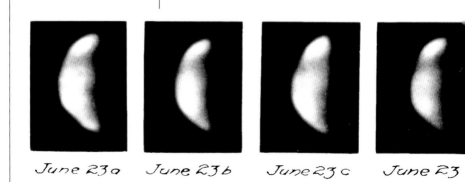

PENETRATING VENUS' CLOUD COVER

Observation of details on the disk of Venus has proven to be very difficult, since only blurred shadows appear near the terminator and many astronomers have had difficulty in interpreting these. One of the first observers who managed to describe them was Francesco Bianchini who devoted himself to the study of the planet in the early eighteenth century and believed that he had determined the planet's rotation period, which he fixed at 42 days 8 hours. He also drew a chart of Venus, dedicating hypothetical seas and lands to the king of Portugal. The existence of an extensive atmosphere surrounding the planet became obvious when Venus transited the Sun in 1874. In fact, on that occasion, the dark outline of the planet's disk just before it started to transit the Sun, appeared to be surrounded by a faintly luminous ring which remained visible until the planet had moved at least one degree away from the Sun, after which it shrank back to its usual half-moon shape. In 1932, Walter S. Adams and T. Dunham discovered a peculiar series of absorption lines in the region of near infrared in the planet's spectrum. From the regularity of the series of lines, typical of certain molecular bands, and from their wavelengths, it could be inferred that these were absorption lines caused by carbon dioxide molecules. The series was also observed in the laboratory and the identification was confirmed: the inference was that above Venus's clouds there is a significantly dense atmosphere in which carbon dioxide predominates.

Meanwhile, the structure of the surface layers of Venus had been clearly reproduced in images taken in ultraviolet light but, despite considerable research by various scientists, it was only in 1960 that Boyer was able to demonstrate that the entire atmosphere rotates around the planet in a retrograde direction, with a period of approximately four days. This phenomenon was subsequently confirmed by spectroscopic observations undertaken in 1965 and, later, by data gathered by interplanetary probes.

Below: The morphology of the structures present on the disk of Venus is revealed by these ultraviolet photographs taken in 1928 by Frank Ross with the Mount Wilson telescope. Despite the evidence provided by these details, it was only in 1960 that Boyer realized that the atmospheric layers rotate around the planet in a retrograde direction, the rotation period being about four days. This has been confirmed by images sent back to Earth by space probes.

June 24 a   June 24 b   June 24 c   June 24 d

EARTH

Above: The Earth as seen by Meteosat 1. The clouds are shown in bright yellow, while the oceans are blue and the continents are brownish-reds.

Physical characteristics

The Earth, together with Mercury, is the largest and the most dense of the terrestrial planets. It is the only one to have a large satellite, the Moon; Mercury and Venus do not have satellites, and Mars has two but they are so small they are completely insignificant in relation to the planet's dimensions. In the case of the giant planets, these have satellites that are equal in size or even larger than the Moon, but these are of negligible mass compared with that of their respective planets. It is necessary to reach the outer limits of the Solar System to find another small planet with a large moon: Pluto, which, with its satellite Charon, forms a "double" planetary system.

The characteristics that make Earth a unique planet is the presence of an atmosphere that is mainly composed of nitrogen with a high percentage of oxygen, and a large quantity of surface water. In addition, Earth has a relatively stable temperature with very small variations between night and daytime temperatures. These are the fundamental requirements for life forms to develop. It is not yet known with certainty whether other living organisms, differing from terrestrial ones, can exist and develop in drastically different conditions, nor is it possible to say if and when it will be possible to give a definitive answer to this question. Obviously, Earth is the planet about which most is known of the interior, surface and atmosphere, despite the fact that it was the last to be

Atmosphere

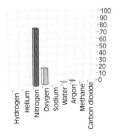

observed "from the outside," by means of artificial satellites. Pictures of Earth taken from space show a panorama of cloudy regions, varying continuously in shape and extension, and brilliant white in color due to their high capacity to reflect sunlight, while the oceans appear as dark blue and the continents green and reddish. The albedo of the thickest clouds can often reach 80%, that of the oceans is only 6%, that of the deserts 35%, while the albedo of regions covered by vegetation does not exceed 20%. Earth's is the only planetary surface in which material in a solid state and in a liquid state have a stable co-existence. Compared with the other terrestrial planets, it has relatively few impact craters. It is also the only one where there is known to be a continual interchange between the water of the oceans and the water vapor contained in the atmosphere.

Shape and rotation

When we talk about the Earth, we think of a celestial body that is spherical in shape. In reality, Earth is slightly flattened at the poles, due to the effect of rotation, and its geographic shape approximates to that of an ellipsoid. The difference between the equatorial radius and the polar radius is only 21 km (13 miles), but this is nevertheless enough to cause precession. In fact, if the Earth did not rotate, the gravitational action of the Moon and the Sun on the "equatorial bulge" would cause the rota-

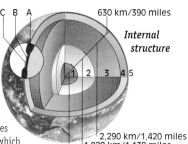

Internal structure

630 km/390 miles
2,290 km/1,420 miles
1,820 km/1,130 miles
1,600 km/1000 miles

1—Inner core
2—Outer core
3—Lower mantle
4—Upper mantle
5—Crust

The crust consists of:
A—Continental crust (40 km/25 miles)
B—Oceanic crust (10 km/6 miles)
C—Ocean (4 km/2.5 miles)

Characteristics of Earth

Mean distance from Sun (AU)	1.00	Mass (g)	5.976 x 10²⁶
Mean distance from Sun (10⁶km)	149.60	Mass (Earth = 1)	1
Orbital period (days)	365.256	Equatorial radius (km)	6.378
Mean orbital velocity (km/s)	29.79	Equatorial radius (Earth = 1)	1
Orbital eccentricity	0.0167	Mean density (g/gm³)	5.52
Apparent mean diameter of Sun	31'59"	Mean density (Earth = 1)	1
Inclination of orbit to ecliptic (°)	0.000	Volume (Earth = 1)	1
Number of satellites	1	Ellipticity*	0.0034

Equatorial surface gravity (m/s²)	9.78
Equatorial surface gravity (Earth = 1)	1
Equatorial escape velocity (km/s)	11.2
Sidereal rotation period at equator	23 hours 56 min 4 sec
Inclination of equator to orbit (°)	23.45

Ellipticity is (Re—Rp)/Re, where Re and Rp are the planet's equatorial and polar radii, respectively.

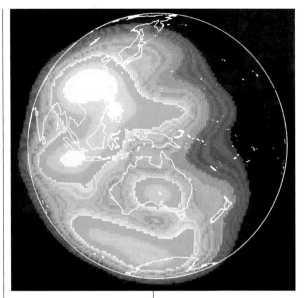

Above: The Earth as seen by the Giotto probe. Cloud formations are indicated by the colors green and blue, above Australia, Asia and the Antarctic.

tional axis to coincide with the perpendicular on the plane of the orbit and therefore make the equator coincide with the ecliptic. But since the Earth does rotate, we have the identical effect to that seen in a spinning top rotating around an axis at an angle to the floor: the axis describes a circle around the vertical and its inclination remains the same so long as the top continues to spin. Today, the Earth's axis of rotation is directed towards a point very close to Polaris, the brightest star in the constellation of Ursa Minor. In 3,000 BC, the polar star was Thuban in the constellation of Draco; 2,000 years ago it was in the constellation of Aries; and in about 13,000 years from now, it will be Vega in the constellation of Lyra, one of the most brilliant stars in the northern sky, which can be observed in summer when it is almost at its zenith.

Gravitational field

The shape of the Earth is described as a geoid. This term describes a solid whose surface is defined by the surfaces of the oceans and, on solid ground by measurements of the force of gravity. The shape of a geoid is

not as regular as that of an ellipsoid, since numerous undulations occur in the former caused by the distribution of the masses beneath the terrestrial surface. A property of the geoid is that at any given point of it, the direction of a plumb line is perpendicular to its surface. Measurements of gravitational force show that heterogeneities exist inside the planet. With the advent of the space era, artificial satellites have provided a very efficient method of measuring the Earth's gravitational field. In fact, the movement of these artificial satellites is influenced by the gravitational attraction of the planet and the small perturbations in orbital elliptical movement have revealed that the upper mantle is a region in which convective currents are active and that Earth is a planet in which dynamic processes are taking place. It has also been discovered that the geoid is not symmetrical in relation to the equatorial plane: the distance of the North Pole from the center of the Earth is approximately 25 meters (82 feet) greater than that of the South Pole, which is why the shape of the Earth has been likened by some to that of a pear. It is now possible to measure the unevenness of the Earth's surface with great precision. Although mountains such as those of the Himalayan range may appear very impressive, in reality elevations and trenches of 10 km (6¼ miles) are very small in relation to the Earth's radius of nearly 6,400 km (4,000 miles): they are, in fact, the equivalent of irregularities of 1 mm or 2 mm (0.039 inches or 0.078 inches) on a sphere with a one-meter (39-inch) radius.

The fact that the Earth's structure is not completely rigid (due to its liquid core, semi-fluid upper mantle, and because the oceans cover two-thirds of its surface) means that even the period of rotation of the Earth is not strictly constant, being subject to fluctuations of more or less than a thousandth of a second: variations, that is, of less than one hundred-millionth of the duration of a day.

Below: The Earth as seen from space.

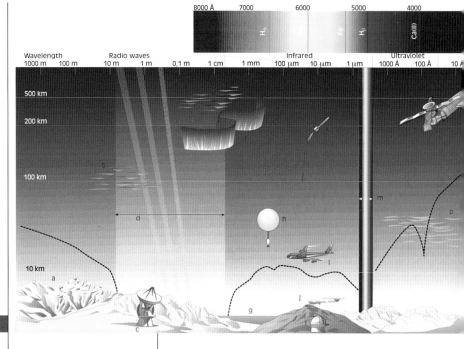

Atmosphere

The composition of Earth's atmosphere is 76% nitrogen and 21% oxygen. Water vapor and argon each account for approximately 1%. Carbon dioxide, methane and ozone are present in quantities that, despite being very small, are anything but insignificant due to the effect they have on the absorbent properties of the atmosphere.

It is thought that when the Earth's primitive atmosphere had just been formed, it had a very different composition compared with the present. Outgassing from the young Earth's crust produced a mixture of water vapor (approximately 80% of the total volume), carbon dioxide (12%), sulfur dioxide (6%) and nitrogen (1 or 2%). Subsequently, most of the water vapor condensed to form the oceans, while most of the carbon dioxide, once it had dissolved in the oceans, contributed to the formation of carbonaceous rocks, such as calcite and dolomitic limestone. Sulfur dioxide is very chemically active and it undergoes a rapid transformation into other sulfur compounds which take it up from the atmosphere. Molecular nitrogen, unlike sulfur dioxide, is chemically inert and has consequently accumulated in the atmosphere to the point where it is the most plentiful component.

Oxygen, the second most plentiful gas and first in order of importance for life, is not a product of outgassing or volcanic activity but is a product of photosynthesis. The latter consists of the formation, under the

action of sunlight, of carbohydrates from carbon dioxide and water with a consequent freeing of oxygen from plants containing chlorophyll. The formation of the diatomic molecule of oxygen is accompanied by the formation of ozone, its molecule consisting of three oxygen atoms.

Atmospheric layers

The atmosphere is divisible into various layers. The lowest and most dense is called the troposphere and contains about 80% of the entire terrestrial atmosphere by mass. It extends to an altitude of 10 km (6¼ miles) and therefore includes all the mountains, even the highest ones. The temperature, greatest on the ground where solar radiation, both visible and infrared, is stored from the atmosphere, falls to 50°C below zero (about −60°F) at the outer edge of the troposphere, while pressure acquires values of approximately one-tenth of those recorded at sea level. The next layer, the tropopause, in which the temperature remains virtually constant, extends up to 25 km (15½ miles), and is surrounded by the ozone layer. Ozone is formed when the molecule of the stable form of oxygen is split by ultraviolet radiation.

The freed atoms can combine with the molecules of oxygen present to form ozone. A continual formation and disassociation of ozone and oxygen molecules takes place and during this process nearly all the solar ultraviolet radiation is absorbed. The ozone layer, stretching from alti-

Left: The solar radiation spectrum (top). Although infrared can be partially observed from Earth through telescopes positioned at high altitudes such as those installed on Mauna Kea in Hawaii, at nearly 5,000 meters (over 16,000 feet) or, better still, from high-altitude aircraft and from balloons, ultraviolet can only be observed by telescopes installed on satellites well above the ozone layer.

a) Mount Everest, 8,888 meters (5.5 miles) above sea level; b) lower ionosphere; c) radio telescope; d) radio waves window; e) high ionosphere; f) aurora borealis; g) sea level; h) aerostatic balloon; i) optical telescope; j) Mauna Loa volcano, 4.17 km (2.6 miles); k) rocket to the Sun, 250 km (150 miles); l) jet, 12 km (7.5 miles); m) optical window; n) Skylab, 435 km (270 miles); p) ozone; q) solar observatory orbiting at altitude of 435 km (340 miles); r) depth of penetration by radiation; s) Mount Rosa, 4,633 meters above sea level (2.87 miles).

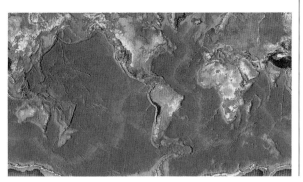

Left: A topographic map of the Earth made by radar altimeter data from NASA's Seasat satellite.

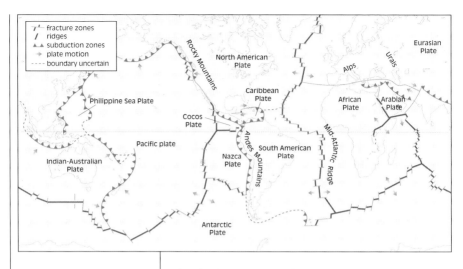

Above: A map of the world showing the various tectonic plates, the Pacific plate being the most extensive. The term plate in this context means an extensive rigid, yet mobile block (speeds of 2 or 3 cm (about one inch) per year have been measured), varying in thickness from 50–250 km (30–150 miles), including the terrestrial crust and the upper part of the mantle. The various trenches, or separations between the different plates are shown: for instance, the mid-Atlantic ridge between the African and Eurasian plates and the north- and south-American plates.

tudes of 25 km up to 30 km (about 15 up to 20 miles), has a very low density. If it were possible to compress it, until it attained the mean sea level atmospheric density, it would become a layer only a few centimeters thick. Its presence is nevertheless of fundamental importance for life on Earth, because without its protection solar ultraviolet radiation would cause irreparable damage to skin and eyes.

This is justification for the alarm caused by the thinning of the ozone layer discovered in recent years and which is thought to have been caused, at least in part, by the input into the atmosphere of chlorine atoms. These are present in various chemical compounds used in air conditioners, refrigerators, jet aircraft engines and by emissions of exhaust fumes resulting from various other industrial activities. A property of chlorine is its very great affinity with the free atoms of oxygen, which is why it tends to subtract them from oxygen molecules, preventing the formation of ozone.

The stratosphere and the ionosphere

The ozone layer itself is part of the stratosphere which extends to an altitude of 50 km (31 miles). The troposphere and the stratosphere contain 99% of the atmosphere's total mass. In the stratosphere, the temperature rises again to nearly 0°C (32°F) due to the process of solar ultraviolet radiation absorption described above. From the stratopause to the mesosphere, which stretches up to an altitude of 100 km (about 60 miles), the temperature falls again, to 80°C below zero (–110°F). Above 80 km (50 miles) the ionosphere begins, so called because molecules are ionized by solar ultraviolet radiation. The ionosphere includes the mesosphere, the thermosphere and the exosphere layers. From the mesopause to the thermosphere the temperature starts to rise again due to the absorption of solar radiation. Here the term temperature, as is also the case with the solar corona, is expressed as an index of the energy of the individual particles. The ionosphere gradually fades into interplanetary space and the pressure that exists in the layers situated at altitudes above 400 km (250 miles) is barely one hundred-millionth-billionth of ground-level pressure. There are several layers in the ionosphere, differing from one another in the density values of the free electrons. This density in turn qualifies the mean properties allowing or impeding the propagation of radio waves of a certain length. The density of electrons in the ionosphere does not increase in a smooth progression with the altitude, but in a series of "jumps" as the altitude increases the density of the gas diminishes and the degree of ionization increases.

The property of the ionosphere totally to reflect radio waves of a given wavelength was exploited from the earliest days of radio transmission to carry out transoceanic transmissions over distances so great that the transmitting and receiving stations were not able to "see" one another, as they were "hidden" from one another by the curvature of the Earth. Nowadays the function of this reflecting layer is fulfilled by radio-

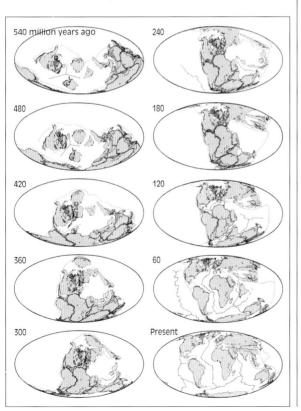

Above: The Earth, when seen from space, looks like a luminous world of blue and white. The presence of an atmosphere composed mainly of nitrogen and oxygen, a vast quantity of surface water and a relatively stable temperature are fundamental factors in the development of life forms.

Left: A schematic representation of the movement of tectonic plates during the course of the Earth's evolution, from 540 million years ago. By approximately 200 million years ago all the land masses above sea level formed a single continent, called Pangea, then gradually changed to the present configuration.

THE CROWDED SKY ABOVE US

In 1957 the International Geophysical Year was marked by the first global research program covering the terrestrial environment, using the most advanced technology of the time. The discovery of the Van Allen radiation belts by the U.S. satellite Explorer 1 *which had been launched in January 1958, was the first step in exploration aimed at the scientific study of the Earth, its atmosphere and environmental resources.*

The Vanguard 1 *satellite, launched in March, 1958, enabled scientists to define the structure of the Earth more accurately; subsequently NASA's* GEOS *and the French satellites* Castor *and* Pollux *carried out geodetic studies, while* Lageos 1 *and* 2, *Japan's* EGP *and the French* Starlette *transmitted data on continental drift. The American* Explorer, *the British* Ariel *and the Italian* San Marco *series of satellites were all involved in measurement of atmospheric density. During the 1960s extensive meteorological research was carried out, with several countries participating, including the United States, China and India; among the satellites involved in this field were* GEOS *and* Meteosat, *which observed the continual evolution of cloud formations. In the early 1980s, when the first space shuttle* Columbia *was launched, a new era of space exploration began. The space shuttles, which took off using traditional rocket launches but landed like normal airplanes, meant that it was now possible for a group of astronauts to carry out varied and important tasks within Earth's orbit. During this same period the* Nimbus *satellite discovered the hole in the ozone layer over the Antarctic and in order to study this more closely NASA built the* UARS *satellite which was inserted into orbit by the* Discovery *space shuttle in 1991. During the 1980s the ambitious project of building an orbiting space station was brought to fruition, with the aim of studying the Earth and serving as a launch pad for exploration of the Solar System. The basic module of the Russian space station* Mir *(Mir means peace), was launched in 1986 and the rest of the space station was assembled directly in orbit. Its altitude was changed several times when required. The* ISS *(International Space Station) was developed by the leading space agencies and provides a large and very well-equipped orbiting laboratory in which gravity-free scientific and technological research is to be carried out as well as an ongoing program with the objective of observing the Earth. The American* Landsat, *French* Spot *and European Space Agency's* ERS *satellites are also playing important roles in terrestrial observation.*

From the early 1990s onwards, Europe and the United States have launched programs to study those environmental changes for which mankind is largely responsible. The new EOS *(Earth Observing System) project aims to gather information on "System Earth," with particular reference to the land masses, oceans, atmospheres and life forms, using large satellites such as* Terra *which was launched in December*

Above: The ERS 1 *satellite during tests carried out at in the Tolouse Interspace Center's anechoic chamber. The satellite was launched in July 1991 from the Kourou space center in French Guiana, with an Ariane rocket.*

1998, followed by Acqua, while other, smaller, long-range, unmanned satellites, tailor-made to specialize in various fields of observation are to be inserted into different orbits, depending on their tasks. The European Space Agency has, however, built a large observational platform, Envisat.

Left: The oceanographic satellite Topex-Poseidon built as a collaborative project by NASA and the French CNES. This satellite was launched on August 10, 1992 from the Kourou space center in French Guyana, using an Ariane 42P rocket. With a launch weight of 2,402 kg (5,295 lb), Topex-Poseidon was inserted into a circular orbit at an altitude of 1,330 km (826 miles) and with an inclination of 66 degrees to the equator.

Below: A drawing of the Envisat platform; at the present time this is the most sophisticated means of monitoring the terrestrial atmosphere, oceans, surfaces and ice formations. It went into orbit in 2001.

communications satellites which are not subject to the variations in density between daytime and nighttime nor to variations in solar activity.

Clouds and climate

On average, clouds always cover approximately 50% of the Earth's surface, even though on a local scale they can vary by up to 100% in the course of a day or even over a few hours. The constant presence of clouds on a planetary scale is due to the continual exchange of water between the oceans and the atmosphere.

A proportion of the solar energy stored by the Earth's surface is irradiated back again into the atmosphere and a proportion is dispersed in space through evaporation of the water and atmospheric convection. This means that the warm upward currents and the cold, downward currents have the effect of transporting heat from the Earth's surface towards space. The small changes in the quantity of cloud, in atmospheric composition and in the properties of the terrestrial surfaces are all causes of climatic variations.

Clouds, formed by water droplets, are continually created by movements of air masses, that is to say, by winds. A vast number of cloud types exist, depending on whether they are produced by large or small-scale air masses and whether rising or falling masses are involved. The highest clouds, at the upper limit of the troposphere, are called cirrus, cirrocumulus and cirrostratus and are all a whitish color. Those at an altitude of 5,000 (16,000 feet) or 6,000 meters (20,000 feet) are called altocumulus (white or grayish) and altostratus (gray or bluish in color) and, finally, the lowest: cumulus; cumulonimbus; stratus; stratocumulus, all of which are very dense and vary widely in shape.

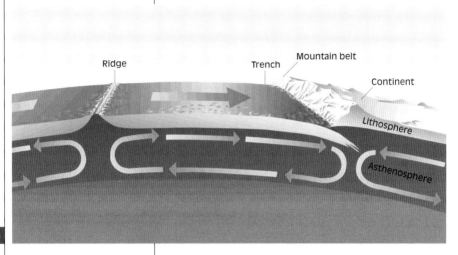

Above: A schematic representation of the terrestrial crust and upper mantle. On the crust are the oceanic depths, mounts, trenches and mountain belts and continents. The whole of the rigid, external crust is called the lithosphere. In the upper mantle, underneath the crust, convective currents occur which cause movement of the plates.

It is well known that variations in temperature corresponding to latitudes and seasons are caused by variations in insolation, resulting from the fact that the Earth's equator is tilted at an angle of approximately 23.5° in relation to the plane of its orbit round the Sun (or ecliptic). The differences in temperature produced by variations in insolation in turn produce differences in pressure which lead to the creation of complex atmospheric circulation and of the regimen of the winds. It is known that there have been climatic variations on Earth over very long periods of time, which may have resulted from tectonic activity causing the continents to shift in such a way that their latitudes have changed drastically during the course of some hundreds of millions of years, or as a consequence of the tilting of the Equator in relation to the ecliptic, which may be caused by variations in the distribution of masses (for example: an increase or decrease in the size of the ice caps). In addition, the largest volcanic eruptions can leave particles of dust suspended in the air, leading to a considerable reduction in insulation over the entire planet. No definitive explanation exists for the various glacial eras, not least because many factors influence climate changes and too many simplifications would have to be introduced in atmospheric modeling to be able to process them mathematically.

Surface

The solid crust and the oceans were formed over 4 billion years ago. Changes in the Earth's crust, the formation of oceanic trenches, moun-

Below, left: The eruption in May 1980 of the Mount St. Helens volcano, Washington State, U.S.A.

Below, right: India, Nepal, Tibet, the Himalayan mountain ranges and the Ganges plain as seen from Apollo 7 at an altitude of approximately 235 km (146 miles) at noon local time.

tain ranges, the continents, all processes that affect the structure of the Earth's surface, can be explained by the theory of plate tectonics. Today we are aware of the existence of extensive submarine mountain ranges, exemplified by the Mid-Oceanic Ridge, running down the entire length of the Atlantic Ocean from north to south. Iceland is situated on the extreme north of this ridge and is a focus of seismic activity, with many active volcanoes, constant features of such oceanic ridges. The basalt lava that erupts along the crests of these ridges is continually forming new oceanic crust which moves slowly away from the crest at speeds varying from 1–10 cm (a few inches) a year. In other regions, especially in the Pacific basin, the oceanic crust is subducted downwards into the mantle; this process results in vast submarine trenches where adjacent "slabs" of the rigid, rocky crust (or lithosphere) converge.

These "slabs" are called tectonic plates and the Earth's crust is made up of 15 of them. The steeply sloping area where one plate is bent under another and descends below it is called the subduction zone. When two plates collide, part of the crust melts and erupts as magma. If the collision zone is under the sea, the magma and sedimentary material can lead to the formation of an island at the edge of the upper plate. An example of this process is provided by the Aleutian Islands in Alaska.

When continental and oceanic plates converge, the subduction zone runs along the edge of the continental plate and the melted material forms a volcanic mountain range. Friction between two tectonic plates

Below: An iridium-rich argillaceous layer in the Apennines. It has been suggested that the layer of dark iridium, trapped between the pale limestone (below) of the late Mesozoic era and grayish limestone (above) of the first Cenozoic era may be the result of asteroid impact. The coin gives an idea of the scale.

causes earthquakes. The convergence of the two continents caused the formation of mountain ranges such as the Alps and the Himalayas. The shape of these mountains has undergone great changes as a result of uplift and erosion.

Continental drift

The discovery of the plates that form the Earth's crust, and of their movements, together with the realization that the edges of the western coast of the African continent and the eastern coast of South America, although separated by the Atlantic Ocean, could fit together like two pieces of a jigsaw puzzle, has resulted in the conclusion that hundreds of millions of years ago the two continents were joined together. By a logical progression of thought, it is believed that all the present continents must once have formed a single, giant super-continent known as Pangea, that emerged from, and was surrounded by, a single ocean, called Panthalassa. Approximately 200 million years ago Pangea started to split up into two continents, Laurasia and Gondwana. North America, Greenland and the greater part of Eurasia probably originated as parts of Laurasia, while South America, Africa, the Antarctic and Australia would once have been part of Gondwana. There is, however, additional, much more convincing evidence for continental drift than the fact that the coastal profiles of two continents could slot into one another. Although they are now separated by the Atlantic Ocean, rock formations dating from the same era and fossils of identical flora and fauna have been discovered in different continents.

A method of tracing the separation of the continents in the past is provided by paleomagnetism, the study of rock magnetism. Terrestrial rocks contain particles of iron. During the rock's formation these particles align themselves towards the axis of the magnetic field in a phenomenon of magnetization that remains stable for billions of years. The study of the magnetization of rocks formed in different eras has revealed that not only does the direction of the terrestrial magnetic pole change over time, but also that the magnetic field is periodically reversed and what is now the magnetic north pole was once the south pole. The curves of polar wandering observed in a given continent do not coincide with those found in another and, since in any given era only a single magnetic north (or south) pole could have existed, the curve of polar wandering had to be unique. The differences found are therefore due to the movement of continental masses. Inverse magnetization found in very ancient rocks indicates that the terrestrial magnetic field was reversed many times.

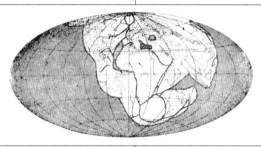

Top: The separation of the two plates corresponding to Egypt and Saudi Arabia. These are moving away from one another and have formed the trench which contains the Red Sea. In this picture, taken in 1966 during an orbital flight, Arabia is on the left and Egypt on the right. The Nile is clearly visible.

Above: The theory of continental drift, proposed by Alfred Wegener in 1912, forms the basis of modern plate tectonics. The Continental plates, which during the Carboniferous era formed a single continent, Pangea, later separated to reach their present positions. Wegener's original explanation has undergone little significant subsequent change.

Above: A picture taken by the Landsat 1 satellite from an altitude of 914 km (567 miles) shows part of eastern Nepal which contains Mount Everest and the upland plain of the Ganges.

Below: The arrows indicate the recently-discovered "Prague basin." With a diameter of nearly 400 km (nearly 250 miles) it may be the largest meteorite crater ever discovered.

Impact Craters

Unlike Mercury, Mars and the Moon, whose surfaces are literally pock-marked by the presence of impact craters, Earth has only a small number because of plate recycling and the effects of erosion by water and wind.

The most famous impact crater to have survived these weathering processes undamaged is to be found in Arizona, it has a diameter of 1,200 meters (4,000 feet), is 170 meters (560 feet) deep and is encircled by a rim that rises approximately 50 meters (about 150 feet) high above the desert plane. On the entire planet, impact craters are in the order of 122 in number, of which 42 are in North America, 31 in Europe, 17 in Australia, 16 in Asia, 11 in Africa and 5 in South America.

Knowledge of the geology of the impact zone and of its history can, in many cases, enable us to calculate the age of the crater. The most recent is at Sihote Alin in Siberia: the fall of the meteorite that produced it was observed on February 12,1947. The Arizona crater is at least 25,000 years old. The oldest known crater is at Sudbury in the Great Lakes region of North America, and its age is estimated to be nearly 2 billion years.

The Earth's interior

Direct knowledge of the Earth's structure is limited to its "skin" alone. Of its 6,370-km (3,960-mile) radius, man has ventured only to depths of

under 4 km (2½ miles), that of the deepest mines; exploratory drilling has reached at most a depth of 10–11 km (6–7 miles).

There is, however, an indirect method that makes it possible to reconstruct the succession of various layers, their temperatures and density and the state of the material: by seismic waves. There are various types of seismic waves; there are longitudinal or P waves when the particles

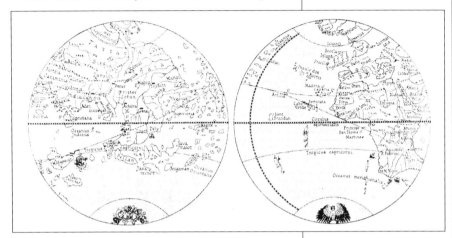

oscillate back and forth in the same direction of propagation as the wave, while there are transverse or S waves when the particles move in a direction that is perpendicular to that of the propagation of the wave. P (or primary) waves and S (or secondary) waves travel through the interior of the Earth and they are therefore also known as internal seismic waves. Only the P waves, however, travel through both solid material and liquid material, whereas S waves cannot travel through liquids. Others travel only near the surface and they are therefore also called surface waves.

From the variation in speed of propagation of the internal waves, it is possible to reconstruct the Earth's internal structure. In the same way as a light ray that travels in a medium of growing density bends regularly as a result of variation in the index of refraction, or undergoes a sudden change in direction when it passes from one medium to another, so seismic waves change their speeds regularly within given layers or, suddenly, on the boundary that separates two layers.

A journey through the Earth

Observation of the movement of seismic waves has revealed a zone situated in the upper mantle, at depths of between approximately 75 km (50 miles) and 250 km (150 miles) in which the velocity of the seismic waves is lower than in the adjacent zones. This is known as the asthenosphere and is probably a zone in which the temperature is near to the melting

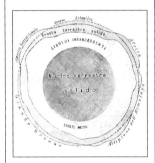

Top: Behaim's globe, drawn in 1492, shows the way in which it was thought the continents were distributed over the terrestrial surface at the time of Colombus's great voyage towards the East Indies.

Above: Earth's interior, in an early twentieth-century illustration.

Below, left: A dynamo consists of a metallic disk rotating inside a magnetic field produced by two permanent magnets (on the left). The magnetic field exerts a force on the free electrons contained in the metallic disk, pushing them towards the center. In this way a potential difference is established between the edge and the center of the disk which produces a current if the circuit is closed (on the right). Certain types of dynamo are constructed so that they generate their own magnetic field, eliminating the permanent magnets. A good conductor of electricity is needed as is a source of mechanical energy to make the dynamo rotate. In a planetary context, a ferrous core represents the conductor and the mechanical energy is provided by the planet's rotation.

point of rocks. It cannot be liquid because S waves travel through it and it must therefore be in a semi-fluid state that accounts for the decrease in velocity of the seismic waves. Above the asthenosphere is the lithosphere where tectonic plates are formed. These plates float on the asthenosphere, and it is the latter's partially molten state that allows the plates to move. If the mantle were completely rigid throughout its depth, continental drift and sea-floor spreading would not have occurred.

From what we have learned so far about the structure of the Earth's interior, we can create a universally acceptable model, according to which, traveling from the center of the planet towards the surface, there is an inner core, probably solid (as indicated by the type of P-wave refraction) with a radius of 1,600 km (1000 miles); an outer, liquid core (which the S waves cannot traverse) 1,820 km (1,130 miles) thick; a lower mantle 2,290 km (1,420 miles) thick, and an upper mantle 630 km (390 miles) thick containing the asthenosphere and the lower part of the lithosphere, where the crust with its tectonic plates is formed.

The magnetic field and the magnetosphere

The Earth's magnetic field is dipolar and has a magnetic north pole and a magnetic south pole, like a normal bar magnet or a long, magnetized iron bar. It is a very weak field: as a comparison, the magnetic field of a

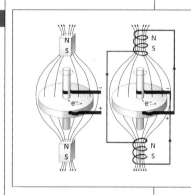
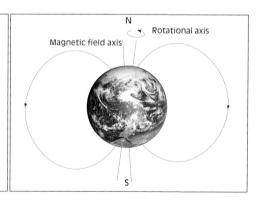

Above, right: Terrestrial rotational axis and magnetic field axis: the two axes form an angle of approximately 11°. The north poles of the two axes are situated in the same direction. In the past, however, the Earth's polarity was reversed and the magnetic south pole was positioned near the geographic north pole.

small magnet has an intensity of several hundred gauss while the terrestrial field measures 0.3 gauss and, as we have already seen, its polar axis is displaced about 11° from its rotation axis. However, despite the fact that it is so weak, its effects are far from negligible. The magnetosphere is a region delineated by the interaction of the solar wind and the Earth's magnetic field. The solar wind flows towards Earth at supersonic speed and is suddenly interrupted by the terrestrial magnetic field, giving rise to the formation of a shock wave; as it is unable to pass into the magnetosphere, it is forced to slide along the surface of separation known as the magnetopause. The magnetosphere is shaped like a comet, with a long, cylindrical tail, that extends over 1,000 terrestrial radii, and a rounded "head." Earth is immersed in the magnetosphere at a distance of approximately 10 terrestrial radii from the top of this "head." The latter is formed where the pressure of the solar wind is equal to the magnetic pressure. On the surface that delineates the magnetosphere, the solar magnetic field's force lines meet up with those of the terrestrial magnetic field. The ionosphere, plasmasphere and the Van Allen radiation belts make up the magnetosphere.

The source of these phenomena is Earth's liquid outer core, traversed by electrical currents that result from the high electrical conductivity of the metals that it contains, especially iron and nickel. The charged particles, circulating freely in this liquid outer core, produce a magnetic field in exactly the same way as an electric current travels along a metal wire. The spinning nucleus of the Earth therefore behaves like a gigantic dynamo, producing the magnetic field that extends for tens of thousands of kilometers into space, interacting with the solar wind. Most of the solar particles cannot penetrate the magnetosphere, but some succeed and are trapped by the Van Allen belts. The lowest belt, about 2,000 km (1,200 miles) from the Earth's surface, contains mainly electrons, while the upper belt, 16,000 km (10,000 miles) from the Earth's surface, contains mainly protons.

Below: The terrestrial magnetosphere. The magnetic field creates a cavity in the solar wind, deflecting the greater part of the solar particles. Those that do manage to filter through the magnetopause are trapped by the terrestrial magnetic field in two regions known as the Van Allen belts, named after the scientist who discovered them.

1. Solar wind
2. Bow shock
3. Magnetopause
4. Force lines of the magnetic field
5. Magnetotail
6. Zone of intense radiation
7. Van Allen belts
8. Geomagnetic equator
9. Magnetic axis

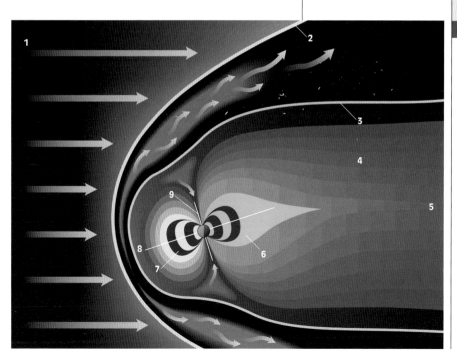

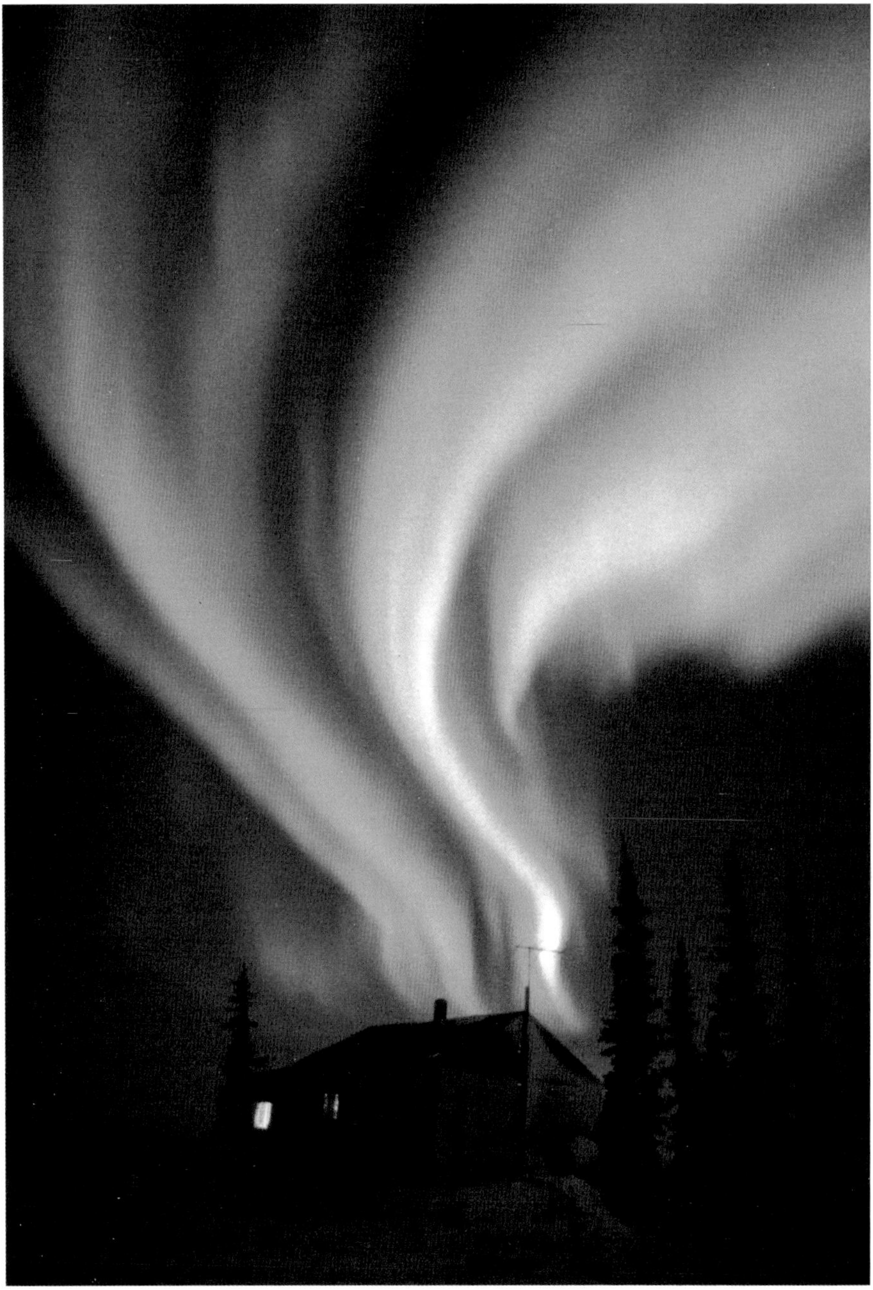

Aurorae

When some violent phenomenon of activity is observed on the surface of the Sun, we are witnessing the production of an excess of particle emissions, not all of which have been trapped by the Van Allen belts. Those particles that manage to escape bombard the Earth's upper atmosphere causing aurora borealis or australis. Because the lines of force of the terrestrial magnetic field are reflected back towards the Earth at the poles, the particles find ingress easier here and this accounts for the greater frequency of aurorae at high northern and southern latitudes. They are more frequent within two approximately circular bands centered respectively around the North Pole and the South Pole. Only exceptional events on the Sun can give rise to aurorae at mid-latitudes. Displays of the Northern or Southern Lights appear as a flickering, colored glow, caused when the energy-charged particles of the solar wind excite the atmospheric oxygen and nitrogen atoms, which then emit their own, characteristic frequencies. Colors change from green to red (oxygen) and from red to green, to a bluish color (nitrogen). There are various types of auroras, divided into two categories and basically classified according to their appearance: they can assume the shape of arches, rays or can look like gigantic curtains.

Facing page, large photograph: Aurora borealis observed in North Dakota on October 10, 1988 and, above, on July 29, 1990.

Some keep their shape for some time, while others change continuously. Photographs of aurora borealis taken simultaneously from two places approximately 100 km (60 miles) apart make it possible to calculate how high they stretch. Most aurorae have their lower edge, or base, at an altitude of 100–115 km (60–70 miles) and the upper edge at approximately 320 km (200 miles): aurorae with a base at altitudes lower than 80 km (50 miles) or higher than 160 km (100 miles) are very rare. If an aurora borealis happens to be observed immediately after the Sun has set, when the lower part of the atmosphere is in shadow but the upper atmosphere is still lit by the Sun, it can be seen to reach altitudes of between 500 km (300 miles) and 750 km (450 miles) and sometimes even 1,000 km (600 miles). As night falls, this is no longer visible and only the normal aurora visible at lower altitudes remains.

Climatic variations

As has already been mentioned, proof exists that in the past Earth experienced major climatic variations. Approximately four and a half billion years ago, when the Sun itself had not yet reached maturity, the climate was much colder than today. However, the terrestrial atmosphere, which contained far more carbon dioxide than it does now, possessed a greater capacity to trap infrared radiation, compensating for the Sun's lower emissions of radiation. Little is known, however, about the Earth's climate during the planet's first billions of years of life.

Data relating to the last 500 million years has been gathered from the fossilized remains of animals and plants and from the processes of rock erosion. A Scot, James Croll, and the Serbian scientist Milutin Milankovic have chosen to explain the alternation of glacial and temperate eras by changes in the geometry of the Earth's orbit, the eccentricity of which varies between an almost circular shape to a more elliptical one during the course of approximately 100,000 years; by polar wandering as a consequence of precession; and also by the lesser effects caused by the attraction of other planets. But apart from the Ice Ages, other minor climatic variations have been detected as having taken place over the last 10,000 years, their causes being numerous and difficult to identify. Large volcanic eruptions, as well as the acceleration in industrial activity, can alter the carbon dioxide content of the atmosphere and can have a considerable effect on climate, perhaps within as short a span as a century.

Left: During the 1920s the Norwegian physicist Kristian Birkeland tried to reproduce the phenomenon of the aurora borealis in his laboratory. He bombarded a magnetized sphere with packets of cathode rays and showed that a ring with the characteristic toroidal shape formed around the sphere, not at all unlike the belts of electrons discovered around the Earth by Van Allen in the late 1950s.

THE MOON

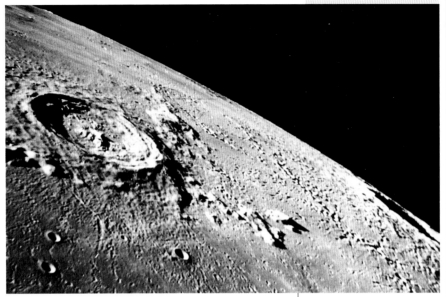

Physical characteristics

The Moon can be described as Mercury's little sister. Mercury's diameter is 1.4 times that of the Earth's natural satellite. Their surfaces are remarkably similar in appearance, pock-marked by craters caused by the impact of meteorites that were not slowed down in their fall, due to the absence of any atmosphere. Their interiors, however, are very different, because the mean density of the Moon is equivalent to 3.35 times the density of water, much lower than that of Mercury. The Moon is, nevertheless, a far-from-insignificant companion to the Earth. Its mass is equivalent to $1/81$ of the Earth's mass, whereas Jupiter's satellites, Saturn and Neptune, as large as, and larger, than Mercury, both have a mass that is less than one-thousandth of their respective planets.

The Moon rotates around the Earth in 29 days, 12 hours and 44 minutes or what we think of as the lapse of time between two successive New Moons. This is an average or mean interval, since disturbances in the Moon's movement caused by the Sun result in this period (which is also known as a "synodic month") varying from month to month. If, instead of having a reference point of a moving body (which the Earth is, since it rotates around the Sun) a point were taken to be "fixed" in space (such as a star that is such a huge distance away that it appears to be fixed in position) the same interval between two new moons constitutes the "sidereal period," equal to 27 days, 7 hours and 43 minutes. The Moon

Above: The Eratosthenes crater and, on the edge, the Copernicus crater as observed by Apollo 17. The former is relatively young, less than 3 billion years old. It is 61 km (38 miles) in diameter and is situated at the southern edge of the Imbrium Sea (or Sea of Showers). The latter is even more recent, being approximately 1 billion years old or less, as indicated by the rays spreading outwards from the crater.

Below: A diagram of the internal structure of the Moon. Like the Earth, it also has a central core, a mantle and a crust. It is possible that the nucleus is iron-rich: its diameter is less than 700 km (approximately 430 miles). The innermost part of the mantle, called the asthenosphere, consists of material that is plastic, (i.e. not rigid) and is surrounded by

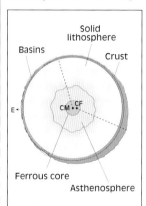

the upper mantle or solid lithosphere. The Moon's crust is thicker on the side furthest from the Earth (approximately 100 km/60 miles thick) whereas the side turned towards our planet is only 60 km thick (about 40 miles). The Moon's center of mass (CM) is shifted by approximately 2 km (1¼ miles) from the center of force (CF): as a result the equipotential surface (the surface whose points are subjected to the same gravitational attraction), lies nearer to the lunar surface that faces the Earth. As a result, the magma that originates far below, in its interior, has greater difficulty in reaching the surface of the hidden hemisphere.

describes an elliptical orbit around the Earth or, more precisely, around the barycenter of the Earth-Moon system. Since the Earth's mass is equal to 81 times that of the Moon, the barycenter actually falls inside the Earth, at 4,670 km (2,900 miles) from its center. The mean distance of the Moon from the Earth is 384,400 km (238,855 miles), which is very small compared with the minimum distance to Venus, the planet next in order of "closeness" to Earth, which is 42 million kilometers (26 million miles) away, or Mars, at a distance of 56 million kilometers (35 million miles). The plane of the lunar orbit is inclined at approximately 5° in relation to the plane of the ecliptic.

When the Moon reaches one of the two points of its orbit known as nodes at which this crosses the plane of the ecliptic, and when it is also a new Moon or a full Moon, the three bodies of the Sun, the Earth and the Moon are in alignment. In the case of a new Moon, the Moon is situated exactly between the Sun and the Earth and there is an eclipse of the Sun; in the case of a full Moon, the Earth comes between the Sun and the Moon: the latter crosses the Earth's shadow and is therefore gradually obscured, but it never becomes completely invisible, because the Sun's light is refracted onto it through the Earth's atmosphere and illuminates it, albeit feebly.

Since the Moon and the Earth are such close neighbors, the gravitational attraction of the latter has made the Moon rotate around its own axis with a period equal to the orbital period. As a result, it is possible to see only one side of the Moon from Earth. We can actually see slightly more than this, for various reasons: the lunar poles are not fixed in space, they precess around the pole of the ecliptic, completing a precessional circle in 18 years, 8 months; the orbit is, moreover, elliptical. The maximum variation from the center of the lunar disk, when these two effects coincide, is 10.5 degrees. This phenomenon and other minor effects mean that, in total, it is possible to observe as much as 59% of the lunar surface. The hidden side of the Moon had remained shrouded in mystery until the Soviet *Luna 3* probe managed to photograph it for the first time in 1959. The photographs sent back by *Luna 3* caused some surprise. For while the side turned towards Earth is covered in large part by extensive plains, known as maria ("seas"), the hidden side has only

Satellite: The Moon

Mean distance from the Earth (km)	384,000
Mean distance in RT	60.21
Sidereal rotation period (days)	27.332
Orbital inclination (°)	18.3–28.6
Orbital eccentricity	0.05
Radius (km)	1,738
Mass (g)	7.35×10^{25}
Mean density (g/cm^3)	3.34

craters and numerous circular structures that bear a resemblance to gigantic craters. Although the reason for this difference is not clear, it is believed that the Earth could be partly responsible. The lunar maria might, for instance, have formed as a result of the impact of celestial bodies traveling between the Earth and the Moon. Moreover, the fact that the same side of the Moon has faced the Earth, probably since a very early phase of the Earth-Moon system, could have had a considerable tidal influence on the shaping of the visible side.

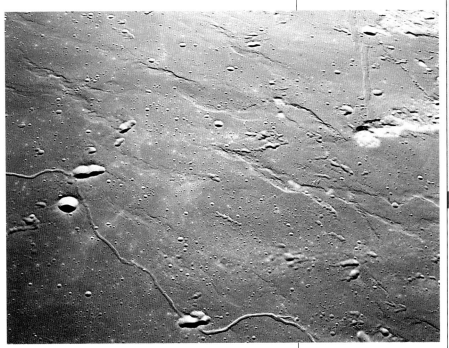

Above: The western section of the Sea of Tranquillity. This was the flat region, with very little unevenness, chosen for the astronauts' landing.

The Surface

The lunar surface consists of two main types of terrain: the highland regions or terrae, covered with craters of all sizes, often very close to one another and sometimes even overlapping, that seem to shine because they reflect more sunlight and, at a lower level, the maria, which are darker and smoother plains.

For a long time there was disagreement over the origin of the craters; nowadays there is no longer any doubt that the overwhelming majority of these and of the large, circular structures known as basins were indeed caused by impacts. Since more recent formations lie on top of one another and cover up the older ones, it is possible to draw geological maps of the whole lunar surface following a stratigraphic classification which makes it possible to reconstruct the past history of the Moon.

The most recent stratigraphic system is called the "Copernican." It includes the more recent craters, among them the Copernicus crater where "rays" that formed following the multi-directional expulsion of material as a result of meteorite impact have remained intact.

The "Herasthostenian" system includes somewhat older craters, in which the rays are no longer visible. The "Imbrian" formations are even older and include the Imbrium basin, a large impact structure that

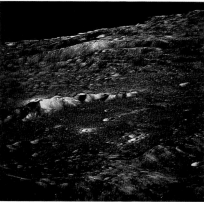

Above, left: The Aristarchus crater, taken from the Apollo 15 *capsule.*

Above, right: A detail of the hidden face of the Moon as observed by the Apollo 10 *capsule.*

was later filled with volcanic material, forming the Mare Imbrium. The "Nectarian" system which includes another large basin, the Nectaris basin, contains four times as many craters as the Imbrian system; this has led to the conclusion that this formation is even older. Finally, the "Pre-Nectarian" system dates far back to the era when the Moon was formed. Recent observations made by the Galileo probe have revealed details of an enormous basin. Judging by its color and characteristics, it would seem to be considerably older than the seas that appear on the visible side of the Moon. Probably the basin was formed by the impact of an asteroid-size object, and consequently produced a large quantity of lava.

The maria and terrae

The maria which were formed from volcanic lava flows (this has been confirmed by analysis of the material brought back to Earth by the *Apollo* mission astronauts) must therefore be even younger. From the reconnaissance of potential landing sites we know that the maria are scattered with fragments of basalt which is also very common in terrestrial volcanic rocks. The age of these maria ranges from 4.3 to 3.1 billion years. Most of the maria were, however, formed between 3.8 and 3.3 billion years ago.

On all the landing sites, astronauts found formations of volcanic origin, spewed out from the depths of the mantle and similar in every respect to the particles formed during eruptions of the volcanoes in the Hawaiian islands. The terrae, or uplands, are composed of ancient rocks with a high calcium and aluminum content. All the lunar terrae must have been formed between 4 billion and 3.8 billion years ago. Also present on the lunar surface are various mountain ranges, the highest peaks of which reach altitudes varying between 3 and 8 km (just under 2 miles and 5 miles). These ranges are generally situated around circular maria.

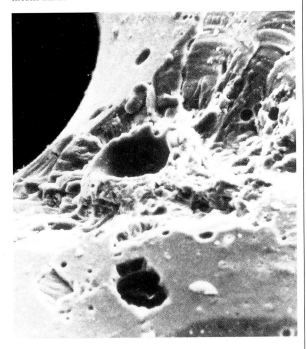

Left: Many lunar samples taken back to Earth are marked by scars caused by impacts from meteorite dust grains which travel at thousands of kilometers per hour. This photograph shows some of these minute craters, less than 1 millimeter in diameter. They act like tiny mirrors which reflect the Sun's light, so that in full sunlight the rocks seem to sparkle. This continuous meteorite bombardment constitutes the sole "meteorological" agent that can change the lunar surface. It is an inefficient agent when compared with those that act on the Earth's surface, taking tens of millions of years to erode 1 mm of the Moon's rocky layer.

Seismology

All information about the interior of the Moon has been gathered from its seismic properties. Since all data on velocity and on the way in which seismic waves travel across the Moon necessitate the use of seismographs placed on its surface, the known data were obtained as a result of exploration by manned missions. Before these, it was only possible to speculate that the Moon's core was much less extensive, in proportion to the planet's radius, than the Earth's core or, at the very least, it had to have a lower iron and nickel content, given the Moon's low mean density, similar to the mean density of the terrestrial mantle. Four types of lunar earthquake have been identified from data collected by terrestrial probes, two of which being the product of forces outside the Moon. The first type was produced by exploding small charges on the surface or by

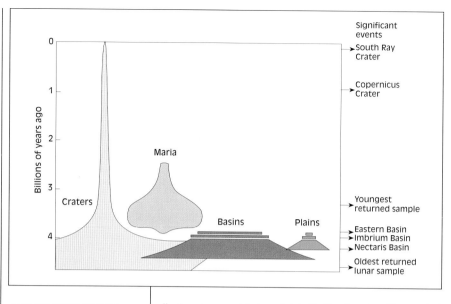

Above: This diagram shows the sequences in which the different lunar rocks were formed. Timescales are shown on the left, stretching from 4½ billion years ago until the present and on the right are shown some of the most significant events which are known to have taken place over this time. The areas covered by this diagram are roughly proportional to the areas of the various formations on the lunar surface. Most of the craters had already been formed 4 billion years ago; most of the maria formed between 3½ and 3 billion years ago. Only craters have continued, and still continue, to form, although in far smaller numbers.

allowing a lunar module to crash into the Moon's surface. The second of these two types is caused by the impact of meteorites. Two other types of earthquake are of natural origin and due to causes within the Moon. A first group originates at shallow depths: these are known as superficial earthquakes; a second group originates at great depths and exhibits a cyclical pattern. Artificially-induced earthquakes have made it possible to ascertain that the lunar crust is approximately 60 km (40 miles) thick at the center of the visible face. While volcanic activity on Earth is mainly due to plate tectonics (the horizontal movement of plates) there is no evidence of the presence of tectonic plates on the Moon.

Earthquakes

Shallow earthquakes originate in the mantle, at depths of 50–300 km (30–180 miles). The seismic energy released by this type of earthquake is a million times lower than that caused by terrestrial earthquakes. It is thought that the most probable cause is the expansion and contraction of the whole Moon due to the cooling of the surface layers and the radioactive interior. During the six years of lunar seismographic activity an average of five seismic events was observed every year. The velocity of the waves in the mantle rises to approximately 8 km/s (5 miles/s) and remains virtually constant to a depth of 1,100 km (680 miles). The second type of natural earthquake is also the most common. These are minor events that occur regularly and which originate at a depth of 800-1,000 km (500-600 miles). During the six-year period that the seismographs were operating, 500 of these were observed, but the energy released was only 1/2,000 of that released by the first type of natural earthquake. This second type of earthquake is also known as a tidal earthquake, because it is clearly associated with the lunar tides caused by the Earth. In fact, events with a given epicenter follow on from one another in a 27.5-day cycle, that is to say the time that the Moon takes to complete an entire revolution around the Earth. Since the distance between the Earth and the Moon varies because of the ellipticity of the orbit, the tidal variations are also linked to the variations in the Moon's distance from Earth.

Internal structure

The lunar mantle has a similar composition to the Earth's upper mantle: minerals present in the greatest quantities are, in fact, olivine and pyroxene. No iron-rich central core has been detected, at least not at depths of up to 1,100 km (about 700 miles).

It is thought that the core has a maximum radius of 680 km (420 miles) and, if anything, the most recent interpretations of seismic data would tend to suggest that the size of the core would turn out to be smaller, with a radius of only 450 km (280 miles) and would account for only 4% of the Moon's volume. Further information on the distribution of material in the interior of the Moon can be deduced from the irregularities that are seen in the movement of probes placed in orbit. These have revealed the existence of regions of denser material close to the surface: these concentrations of matter are referred to as "mascons."

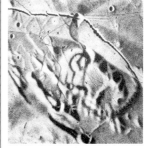

Above: On this photograph of the Huygens rill taken from the Paris Observatory, an early-twentieth century lunar cartographer, J.N. Krieger, superimposed details noted during the course of 15 nights' observations.

Below: Two pictures of the edge of a crater and of the adjacent surface. The second is an enlargement of the area enclosed in the outlined square. The concentric lines around the rim can be very clearly seen.

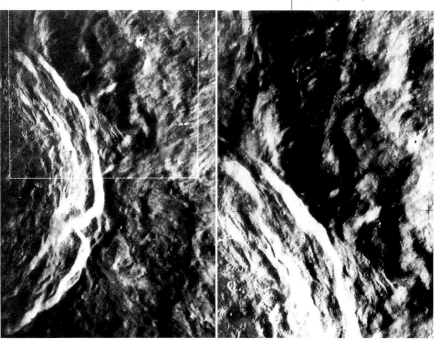

Above: In the mountainous area of the Moon, craters of varying ages are superimposed in a chaotic manner.

Below: The Lunar Apennines. To the left the Alpine Valley is visible, possibly created by a meteorite crashing into the lunar surface.

THE ENIGMA OF THE CRATERS

Of all the Moon's surface structures, its craters are its most typical feature and there has been much debate as to their origin. Galileo recorded them when he chose the Moon as one of the first objects for observation with his telescope, and the most notable of them were given names in Riccioli's Almagestum Novum published in 1651. We still refer to them today as Plato, Aristotle, Copernicus and so on. The hypothesis that they could have been produced by the impact of meteorites was for a long time advanced against that of an endogenous origin, but no satisfactory solution was found to the problem. The theory that eventually came to be most widely accepted was based on observation of the resemblance of the lunar maria to certain types of crater which are circular in shape, containing a vast flat floor, sometimes without any sign of a central elevation. This suggested that they were of exogenous origin, presupposing that they were caused by impacts on the lunar surface when it was already formed by bodies of considerable size: the collision would have liquefied the rocks that now form the "sea-beds." This was the explanation given for the difference between the "highlands" (the most uneven areas of the Moon's surface which are completely covered with craters that intersect and overlap other craters which in turn provide evidence of its most ancient appearance) and the maria. It was already clear that most of the agents that contribute to shaping the Earth's surface are absent on the Moon and that its appearance must have remained almost unchanged for very long periods of time; the origin of the lunar craters had, therefore, to date back to a very remote past.

If a probe transits regions that are more dense than the mean density, its movement accelerates and then slows down when it transits regions of lower density. In general, these mass concentrations coincide with the maria, whereas the Copernicus crater and many basins that are situated on the face that is hidden from Earth coincide with regions of lower density.

Magnetism

The first Soviet probes launched in 1959 had already shown that the Moon has no perceptible general magnetic field. There are, however, regions that show traces of magnetization: from the age of the rocks it

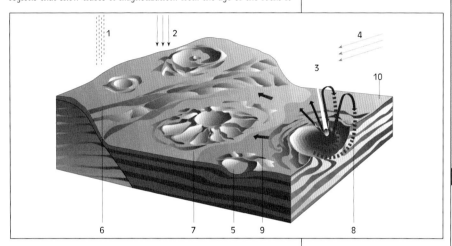

can be established that 3 billion years ago the Moon possessed a magnetic field, which has since disappeared.

Atmosphere

The Moon's mass is too small to be capable of retaining a stable atmosphere. Moreover, the volatile elements that are at temperatures of approximately 130°C (266°F) on the side illuminated by the Sun, reach thermal agitation velocities that are equal or superior to the Moon's escape velocity which is approximately 2.4 km/s (1.5 miles/s).

Since the thermal agitation velocity of the atoms is actually a mean value and many other particles have higher values, this explains how, during the course of hundreds of millions of years, the Moon has lost all its atmosphere. In 1988, however, traces of sodium and potassium vapors were discovered above the lunar surface, indicating that the Moon does possess an atmosphere, albeit an extremely rarefied one, that extends like a long, straight tail towards the Sun up to an altitude of 7,000 km (4,400 miles) above its surface. On the dark side the density is even lower, but the gas extends up to at least 21,000 km (13,000 miles) above the lunar

Above: A schematic representation of erosion agents on the lunar surface:
1) Micrometeorites
2) Cosmic rays
3) Meteorites
4) Solar wind.
The fall of a large meteorite caused the exposure of basalt layers (7), the formation of rays (10) of varying composition above the regolith (6), seismic shock waves (8) and deep seismic waves (9) and small craters (5).

Above: A photograph of the Moon in the last quarter, taken in 1863 by Henry Draper, the first photographer to obtain high quality images of Earth's satellite.

Below: A schematic drawing which summarizes the geological history of the Moon.
A. pre-Nectarian period, from 4.6 to 4.1 billion years ago, when Earth's satellite was formed.
B. Nectarian period, from 4.1 to 3.9 billion years ago, when craters and the Nectaris basin were formed.
C. Imbrian period, from 3.9 to 3.2 billion years ago, when the Imbrium and Eastern Basin were formed.
D. Erastosthenian period, from 3.2 to 1 billion years ago, when the Erastosthenes Crater was formed.
E. Copernican period, from 1 billion years ago to the present day, during which the Copernicus Crater was formed.

surface. Nevertheless, even in the densest areas, the lunar atmosphere's density is only approximately ten atoms per cubic centimeter, compared with the density of tens of thousands of atoms per cubic centimeter of the terrestrial ionosphere.

Origin and evolution

Ideas as to the Moon's origins have changed drastically in less than 20 years, due to the accumulation of results from studies on lunar rocks. Before and during the series of Moon landings by the *Apollo* capsules, there were three hypotheses proposed to explain the origin of the Moon. The "fission" hypothesis supposes that the Moon was once an integral part of the Earth and that it later became separated from it in some way: for example, as a result of the "tides" raised in the Earth by the Sun. Supporters of this theory cited the fact that the Moon's mean density is similar to that of the terrestrial mantle as corroborative evidence. Furthermore, the velocity at which the Moon is moving away from the Earth is compatible with the idea that 4.5 billion years ago they were very close. However, the theory that this separation came about as a result of tidal factors has been shown to be untenable for various reasons. The other hypothesis is that of "capture" according to which the Earth and the Moon were formed independently as separate members of the Solar System and that at a given moment, by a process as yet unknown, the Moon was "captured" by Earth and forced to follow an orbit around it. This hypothesis explains the differences in density between the Earth and the Moon. However, if this capture theory were to be proved, the two bodies would have to have had a very small relative velocity and, consequently, they would have to have been formed at approximately the same distance from the Sun. But, were this the case, they would also have had the same chemical composition. The third hypothesis suggests that the Earth and the Moon were formed together, but this does not explain why the Earth is so much richer in iron than the Moon. The Moon would, moreover, have to have been formed on the equatorial plane of the Earth. In conclusion, none of these

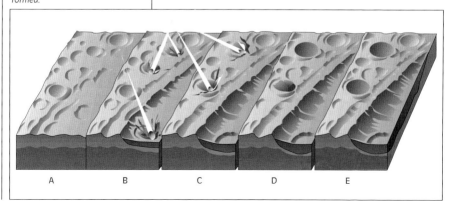

three hypotheses can account for all the observable facts concerning the Earth-Moon system.

Large impact hypothesis

A new hypothesis that has emerged as a result of direct observation, is also known as the "large impact" hypothesis and was suggested by analysis of the possible distribution of the solid agglomerates (or proto-planets) of the solar nebula. Apart from a few proto-planets of large

dimensions, it is probable that there was once a far greater number of these and that these bodies were much smaller in size. In the final stages of the formation of the Solar System a large number of these minor objects collected together to form the planets after violent collisions. According to this theory, the proto-Earth would have been hit by a proto-planet of a size comparable with that of Mars. As a result of the collision, vaporized mantle debris would have been ejected at a high temperature, containing not only material from the terrestrial mantle, but also material from the projectile; but not from the Earth's core. An objection to this hypothesis opposes it with the fact that the fragments from a similar impact would have obeyed Kepler's laws and have moved either along open hyperbolic orbits, thus escaping from terrestrial attraction, or along closed elliptical orbits that would have brought them back to their point of departure, that is, back to the Earth. However, in a high velocity collision, a large part of the projectile and an equal quantity of the body struck would have vaporized and as the vapor expanded it would have reached altitudes equivalent to many terrestrial radii before condensing. Finally, once the vapor had condensed, it would have entered an orbit that did not intersect with that of the Earth, in accordance with Kepler's laws.

Another possibility is that the violence of the impact may have changed the trajectory of the fragments which could then have settled

Above, left: The visible side of the Moon as photographed by the Apollo 16 astronauts in 1972.

Above, right: A picture of the hidden face of the moon. The morphology of the two sides differs considerably: the extensive plains, called maria*, which occur on the visible side, do not occur on the hidden side.*

Above: With the Sun low on the horizon, the long shadows projected by the lunar mountains enabled Galileo to calculate their height. In this photograph the Plato circle, in the Lunar Alps, is visible, with the flat plateau and the Sea of Showers, punctuated by high mountains.
Below, left: The Moon in the last quarter.
Below, right: The Moon in the first quarter.

on a stable orbit around the Earth. Colliding with one another, these fragments might have eventually condensed to form one single body: the Moon. The material from the mantle and from part of the projectile which were involved in its formation would explain the similarities and differences of the Moon's chemical composition compared with the Earth's mantle.

Age of the Moon

Isotopes present in lunar rocks have made it possible to establish that the origin of the Moon dates back 4.5 billion years, which means that it is indeed about the same age as the Earth. So far, it has not been possible with methods currently available to identify these stretches of time more accurately than to the nearest 50 million years. In other words, it is impossible to say whether the Earth and the Moon are contemporaries: one could be up to 50 million years older than the other.

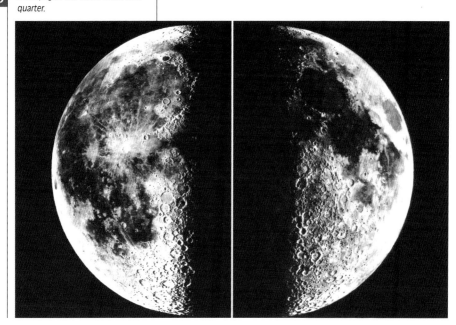

THE CONQUEST OF THE MOON

Lunar exploration was started by the Soviets who were the first to make a probe land on its surface (September 12, 1959) and the first to explore the surface with a vehicle, the Lunokhod, which could move around—helped by the images that two frontally-mounted television cameras transmitted to terrestrial stations—and was steered by remote control from Earth.

NASA took the same approach but it was not until July 28, 1964 that the first US probe to reach the lunar surface, Ranger-7, was launched. The subsequent exploratory missions using the automatic probes Lunar Orbiter and Surveyor all had the same objective: to prepare for a manned trip to the Moon, as promised by President John F. Kennedy in 1961, in order to demonstrate the technological and military superiority of the United States over that of the Soviet Union and to erase memories of the humiliation caused when the Soviets launched the first artificial satellite, Sputnik 1 and achieved the first manned orbital flight, by Yuri Gagarin. On May 31, 1966, the Surveyor 1 probe was launched from Cape Canaveral carrying a tiny space robot which resembled a spider, and on June 2 its legs came to rest on the sandy regolith of the Sea of Storms. This mission was a great success, not least because Surveyor's television cameras were a great improvement over those in Lunokhod and made it possible to acquire very detailed images. On July 16, 1969, the Saturn 5 vector rocket rose from Cape Canaveral's launch pad carrying the Apollo 11 capsule towards the Moon. During the night of July 20–21, Neil Armstrong and Edwin Aldrin touched down on the Sea of Tranquillity in the lander or LEM (Lunar Excursion Module) "Eagle" after separating from the Command

Above: The astronaut Harrison Schmitt beside a huge lunar rock

Below: A photograph taken from on board the Apollo 11 capsule which shows the LEM (Lunar Excursion Module) moving away into the distance prior to landing on the Moon.

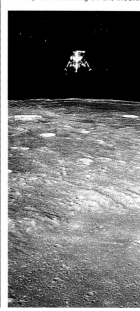

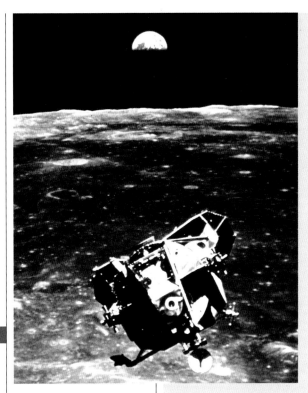

Module which remained in orbit with the third astronaut, Michael Collins, on board. After their visit to the Moon—which lasted 21 hours 36 minutes, during which Armstrong and Aldrin took a Moon walk lasting 2 hours 31 minutes—the three astronauts headed back to Earth, taking with them 21.4 kg (47 lb 3 oz) of lunar rocks and a thin metallic sheet which they had spread out during their Moon walk to gather particles from the solar wind. They left a laser light reflector behind on the Moon to enable distance measurements to be taken from Earth and a seismometer to transmit evidence, if any, of seismic activity. The Apollo 11 mission ended with the recovery of the space capsule and the three astronauts inside it from the Pacific Ocean. This first mission was followed by five other American Moon landings by astronauts, at different locations. During these later missions, a vehicle capable of carrying two astronauts, the Lunar Rover, was also used to explore larger areas of the lunar surface. The last manned flight to the Moon took place in December 1972.

Above: The LEM photographed from on board the command module of the Apollo 11 capsule as it draws near prior to re-attachment after completing its lunar exploration.

Right: The US astronaut Harrison Schmitt at the controls of the Lunar Rover during the Apollo 17 mission.

WATER IN THE CRATERS

Twenty years on, the United States resumed lunar exploration when the small probe, Clementine was launched in January 1994, under the aegis of the Defense Department, to test new technology. On board there was also a group of instruments which, during 71 days of reconnaissance, compiled a relief map of the entire lunar globe. Most sensationally, the probe managed to detect that ice made of water could be hidden in the Aitken crater. The Pentagon announced this discovery at a press conference in December 1996.

After NASA sent the Lunar Prospector into lunar orbit on January 7, 1998 more data was gathered during the months that followed. The new data indicated the presence of ice in impact craters in the moon's polar regions. It has been speculated that the water was brought there by comets in the remote past. Water is the most important raw material necessary for the colonization of space.

Above: One of NASA's Ranger probes during preparations for its launch from Cape Canaveral with an Atlas-Agena B rocket. These probes weighed 369 kg (812 lb) and were equipped with a photographic system manufactured by RCA Astro-Electronics Division which made it possible to obtain images with a resolution of approximately 3 meters (10 feet).

Left: The launch from Cape Canaveral on July 16, 1969 by the Saturn 5 rocket with the Apollo 11 shuttle carrying the first lunar explorers: Neil Armstrong, Edwin "Buzz" Aldrin and Michael Collins. Armstrong and Aldrin's Moon landing took place on July 20 and their stay on the Moon lasted 21 hours 36 minutes. During the time they spent on the lunar surface, the two astronauts' Moon walk lasted 2 hours 31 minutes. Their return to Earth took place on July 24, 1969.

MARS ♂

Above: Mars as seen by the Viking 1 probe. Almost in the center and near the terminator can be seen the great impact basin of Argyre, its shiny interior possibly due to a covering of ice or frost. The south pole is to the bottom right. North of Argyre can be seen Valles Marineris.

Below: The comparative sizes of Earth and Mars and a comparison of their axial inclination in relation to the perpendicular of their orbital plane.

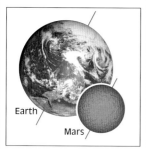

Physical characteristics

Mars is the second closest planet to ours, after Venus, and has long been an object of fascination to the inhabitants of the Earth. Its spectacular appearance and reddish hue make it easily recognizable. When observed through a telescope, it is possible to make out some details, such as the polar caps which are extensive during the Martian winter but virtually absent during the summer, as well as the presence of an atmosphere that varies in transparency depending upon meteorological conditions. The appearance and disappearance of widespread clouds and of certain seasonal changes had led observers to think that there might be vegetation on the planet, a hypothesis subsequently disproved by the space probes which revealed its barren and inhospitable landscape. Certain features had also encouraged people to believe that there was a greater similarity between Mars and the Earth than is actually the case. One statistic that encouraged this misconception is the duration of the Martian day, which is almost the same as the terrestrial day at 24 hours 37 minutes 22 seconds, as well as the existence of seasons that change in a remarkably similar way to those on Earth. In addition, the equator on Mars is inclined at an angle of 25° to the orbital plane, close to the Earth's orbital inclination which is 23.5°.

The Martian year lasts 687 terrestrial days, the equivalent of 23 terrestrial months: hence every season on Mars lasts, on average, approximately 5.75 times one terrestrial month. The orbit of Mars is, however, much more elliptical than that of the Earth and its distance from the Sun varies from 294 million kilometers (183 million miles) at aphelion to 206 million kilometers (128 million miles) at perihelion. Consequently, the orbital velocity of Mars at aphelion is significantly lower than that at perihelion. As a result the length of the Martian seasons is also influenced: in the northern hemisphere spring lasts for 194 Martian days, summer for 177, autumn for 142 and winter 156. In common with the Earth, Mars is at perihelion when it is winter in the northern hemisphere. The diameter of Mars, which measures 6,794 km (4,222 miles) at the equator and 6,759 km (4,200 miles) at the poles, places it half-way between the Moon and Earth as to size, as do certain of its other features. For instance, Mars does possess an atmosphere, albeit a very tenuous one, with a much lower ground-level pressure than Earth's. The Martian surface also has impact craters scattered over it, as well as volcanoes, plains, deep canyons and numerous dry river beds, evidence of the abundant presence of water in a far-off past.

Observations from Earth are often severely hampered by interference due to turbulence in both the terrestrial and Martian atmospheres. Only in 1965, when the first probes were sent towards Mars, was it possible to obtain clear images. Today, after the launching of nearly thirty Martian probes, it has been possible to photograph the entire surface of the planet, showing details down to a few meters; in addition, the two U.S. *Viking* and the Mars Pathfinder probes managed to land in two different sites on the planet, and sent the information they gathered directly back to Earth.

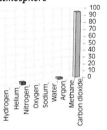

Internal structure

100 km/62 miles
1,800 km/1,100 miles
1,500 km/ 900 miles

1—Core of iron and ferrous compounds
2—Mantle of silicates
3—Crust

Atmosphere

Characteristics of Mars

Mean distance from Sun (AU)	1.52	Mass (g)	6.421×10^{26}
Mean distance from Sun (10^6km)	227.94	Mass (Earth = 1)	0.107
Orbital period (days)	686.98	Equatorial radius (km)	3,393
Mean orbital velocity (km/s)	24.13	Equatorial radius (Earth = 1)	0.532
Orbital eccentricity	0.0934	Mean density (g/cm^3)	3.95
Apparent mean diameter of Sun	21'	Mean density (Earth = 1)	0.72
Inclination of orbit to ecliptic (°)	1.850	Volume (Earth = 1)	0.150
Number of satellites	2	Ellipticity*	0.0052

Equatorial surface gravity (m/s^2)	3.72
Equatorial surface gravity (Earth = 1)	0.38
Equatorial escape velocity (km/s)	5
Sidereal rotation period at equator	24 h 37 min 22 sec
Inclination of equator to orbit (°)	25.19

*Ellipticity is (Re—Rp)/Re, where Re and Rp are the planet's equatorial and polar radii, respectively.

Above: The south pole of Mars as observed from Earth at different times to demonstrate the rotation of the planet.

Right: An image of Mars taken by the Hubble space telescope with the Wide Field Planetary Camera, when Mars was 85 million kilometers (52 million miles) from the Earth. The image resolution is 50 km (31 miles).

Mars is far from a perfect sphere in shape and its divergence from the spherical is far more marked than is the case with Earth. The crust of the northern hemisphere is thinner than that of the southern hemisphere.

Atmosphere

The average ground temperature is 40°C (40°F) below zero, rising to nearly 30°C (86°F) above zero at an altitude of 200 km (124 miles), as a result of ultraviolet solar radiation and X-rays. Pressure at ground level, equivalent to 0.0065 atmospheres, falls to reach a value that is a billion times lower at an altitude of 160 km (100 miles). With such a low ground pressure (comparable to terrestrial pressure at an altitude of 30 km (about 20 miles) and low temperatures, water is unstable and freezes. Minute particles of sand are taken up into suspension and form the clouds of sand that are one of the main features of the Martian environment. The daily variation in temperature on Mars is much greater than on Earth, where the atmosphere acts as an efficient thermal regulator. Within the space of half a day in the Martian summer the temperature can reach values of approximately 10°C (50°F) above zero, only to fall to approximately 70°C (90°F) below zero during the night. The polar temperatures are far lower, at around 120°C (180°F) below zero. Despite being extremely forbidding in the "temperate" zones and on the equator, the climatic conditions on Mars are, therefore, not unlike those in terrestrial polar regions or those experienced at the summit of Mount Everest. Although the Martian landscape can resemble the Sahara desert, the most striking impression await-

ing any future astronauts who might land on the planet would be the color of the sky: the rarefied Martian atmosphere is not capable of scattering the blue and ultraviolet solar radiation which make the Earth's sky blue, and the dust in suspension on Mars also helps to create a yellowish-pink sky. The composition of the Martian atmosphere consists of 95% carbon dioxide, 2.7% nitrogen, 1.6% argon and even smaller percentages of oxygen, carbon monoxide and water vapor. In this respect, the atmosphere resembles that of Venus rather than the terrestrial atmosphere.

Climate

The low density of the atmosphere prevents efficient heat circulation though wind and temperatures differ greatly from one region to another. Another factor that influences climatic variations is the water vapor and the carbon dioxide in the atmosphere which freeze to form ice and dry

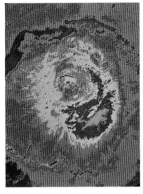

Above: A false-color image of the caldera of the Olympus Mons volcano, taken by Viking 1. *Red indicates those zones covered by older, smoother lava. Some lava flows are not pockmarked with craters, indicating that they are younger.*

Left: The caldera of Olympus Mons.

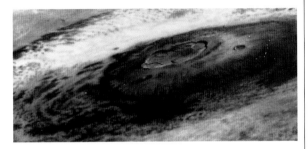

ice over the polar caps, and release latent heat. The dust storms which ravage the entire planet also have drastic effects on the climate.

While the *Viking* probes and more recent probes were carrying out reconnaissance, they observed various types of Martian clouds: some were similar to cirrus cloud, others were obviously cyclonic and yet another type was observed around the mountainous regions, usually accompanied by morning mist, when the frozen ground is warmed by the Sun. In addition to the water vapor clouds, carbon dioxide clouds were also recorded. The dust storms which are a feature of Mars and which often stretch all over the planet, could result from the winds caused by great disparities in temperature, such as those seen near the polar caps, between the zones in which the ice has retreated and those which are still frozen. Their existence has been known for some time and during a dust storm it is often impossible to make out details on the Martian surface.

Surface

The geology of Mars and especially its surface, analysis of which has enabled us to reconstruct the formation and history of the planet, is very different from the Earth's. Mars does not appear to have tectonic plates whereas on Earth these have played, and still play, a crucial role in terrestrial geological processes. Furthermore, water in its liquid state is unstable on the Martian surface, despite the fact that in a remote past the

Above: An image of Mars in false colors which exaggerate the color differentiation. The bright regions, such as the atmospheric mists, the ice-covered surfaces and the pale desert zones are represented by turquoise, white and yellow. The darker zones are shown in dark red and blue. A wide bluish-white band indicates the atmospheric haze which stretches from the volcanoes towards the north terminator.

situation was probably different, judging from the evidence of riverbeds.

The striking difference between the southern hemisphere which is pockmarked by many craters and the northern hemisphere with its wide plains, is something for which it is difficult to find an explanation. If each type of feature on the Martian surface is subjected to scrutiny, it becomes obvious that the easiest of these to observe from Earth, even with an unsophisticated telescope, are the polar caps which expand and contract according to the seasons. It needs to be remembered that Mars, like Earth, is at perihelion when it is winter in the northern hemisphere and summer in the southern hemisphere. Therefore, as a result of its very elliptical orbit, the summers in the southern hemisphere are short and hot and the winters are colder and longer. Around the caps, covered in ice and dry ice as far as latitudes lying at approximately 80°, there are stratified deposits, several kilometers deep, consisting of dust and ice.

This stratification is clearly visible on the slopes of the valley walls. The north pole is surrounded by a formation of dunes which form a ring. Around the south pole, the formations are more irregular. Between latitudes of 80° and 60° north there are extensive plains and occasional craters, while at the same latitudes in the southern hemisphere the ground is almost completely peppered with craters, with only the occasional plain between them. Stretching out between 60° north and the equator are the great plains of Vastitas Borealis, Arcadia Planitia, Aci-

dalia Planitia, and Chryse Planitia where the *Viking 1* probe landed, and sent back the first images of a desolate and barren landscape to Earth. Other notable formations are Utopia Planitia, Syrtis Major Planitia and a region covered in craters, of which those named after Cassini and Antoniadi are the most impressive in size. This same region is surrounded by a series of hills with flat tops and steep slopes, a formation for which a descriptive Spanish word is used: *mesa*, meaning "table." Between the equator and the latitude of 60° south there are also numerous craters;

Above: A detail of the region known as the "Chandelier." In this area the terrain is very uneven and forms a complex series of inter-linked ditches or long, narrow depressions.

Below: A detail of Valles Marineris, observed by Viking 1 from an altitude of 2,000 km (1,243 miles). The area shown measures 300 x 400 km (185 x 250 miles).

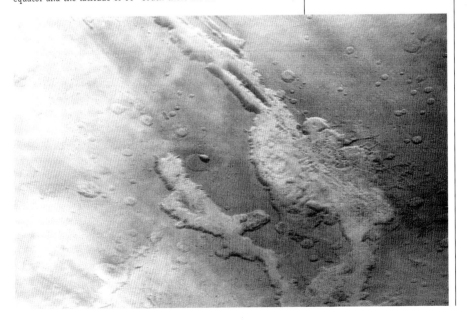

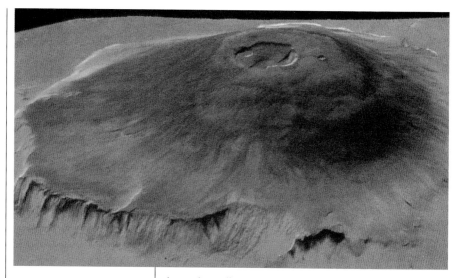

Above: The enormous mouth of the crater of the Olympus Mons volcano which towers to an altitude of 21 km (13 miles) (nearly three times the height of Mount Everest) above the surface of Mars. This is a computer-generated image which emphasizes its vertical dimensions.

Right: A picture of the Martian landscape taken by the Rover Sojourner. The planet's surface seems to show signs of a far-off past with plenty of water which smoothed out the plains and carved out the canyons.

the very long Valles Marineris canyon, a huge circular basin called Argyre Planitia and, further to the east, another vast, circular basin, Hellas Planitia.

Craters

In light of information gathered during the most recent observations carried out by space probes, the craters on Mars are either of volcanic origin, or caused by meteorite impact. The latter can vary from small holes to *fossae*, ditches or long, narrow depressions 200 km (124 miles) or more in diameter. As with lunar craters, Martian craters are usually

ASTRONOMERS AND MARS

The feature now known as Syrtis Major was identified for the first time by Christiaan Huygens in 1659. From the movement of Syrtis Major, this astronomer estimated that the planet achieves a complete turn on its axis once in 24 hours. In a drawing of his dated 1672 the southern polar cap is visible.

Various markings, including the white polar caps, were also observed on Mars in 1666 by Gian Domenico Cassini, who estimated the rotational period of the planet to be 24 hours 40 minutes, very close to the modern value of 24 hours 37 minutes 22 seconds. Giacomo Filippo Maraldi spotted the cap at the north pole and noted changes in the polar caps which were later studied in detail by Wilhelm Herschel at the end of the eighteenth century.

Herschel was the first astronomer to work out the planet's axial inclination in relation to the perpendicular on the orbital plane, calculating it at approximately 30°, whereas the modern value is 25°. It should be remembered that when Mars turns its northern hemisphere towards the Sun, the planet is very near to aphelion, whereas when it has its southern hemisphere facing the Sun it is at perihelion; as a result the heat received from the Sun varies to a greater extent in the southern hemisphere than in the planet's northern hemisphere. As a result of his studies, Herschel discovered the seasonal character of the changes in the polar caps which have maximum extension at the beginning of the Martian spring, minimum extension at the end of the summer. He pointed out for the first time the analogy with the terrestrial polar ice caps. He did not devote much time to the dark areas, which appeared indistinct to him and which he interpreted as clouds, but he did note that they grew fainter when the planet's rotation displaced them towards the lower part of the planet's disk and concluded that it was the presence of an atmosphere that brought about these absorption effects. What was possibly one of the most unexpected discoveries in the history of Mars was made by Schiaparelli in 1877, during the great opposition, when he discovered the famous Martian canals during his lengthy study of the planet's surface.

On this page: Pictures of Mars, all showing Syrtis Major on the meridian.

Above, from the top: A drawing by Frederik Kaiser dated 1864; a drawing by Schiaparelli dated 1879 and another, by Antoniadi, dating from 1909.

Below, left: A drawing by Percival Lowell dated 1894 and a photograph taken by E.C. Slipher in 1941 from the Flagstaff Observatory, Arizona. The north pole of the planet, as is usual in astronomical photographs, is at the bottom. The shadings have a diagrammatic appearance in Schiaparelli's drawing and numerous, very thin channels (the "canals") are present in Lowell's drawing, but missing in Antoniadi's.

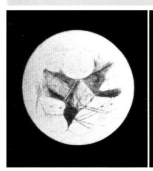

Right: This picture was taken by the Rover Sojourner. Yogi, a Martian rock approximately 1 meter (3 feet) in size is clearly visible; this was the second rock examined by the small robot and by its APXS (alpha proton X-ray spectrometer).

Below: Another detail of the Martian terrain. This photo, taken on September 17, 1997, clearly shows the rock that was given the name of Chimp which has a large crack and is covered by some darker incrustations.

almost circular, surrounded by rims while their floors are at a lower level than the mean ground level, and the surrounding area is scattered with debris expelled at the moment of impact. Among the largest and best preserved are the Hellas basin which is 2,000 km (1,200 miles) in diameter, and the Agyre basin which is 1,200 km (750 miles) in diameter, and is greater than the distance between Chicago and Miami.

As is the case with the Moon, planetary scientists think that the majority of impacts on Mars occurred at least 3.8 billion years ago; it would seem to follow, therefore, that the terrains where craters are most numerous must be the most ancient. The shape of some Martian craters differs from their Lunar counterparts because the expelled material is dispersed in a regular, even pattern around the crater, resembling petal-like lobes and forming one or more rings. This regular formation is explained by the presence of water which saturated the expelled material, forming a muddy mixture that flowed over the ground all around the crater.

Volcanism

The most spectacular volcanoes on Mars are the complex of the Tharsis Montes, Elysium Mons and Olympus Mons. One of the Tharsis Montes, Ausia Mons, is situated approximately 10° below the equator, extends over 400 km (250 miles) and towers to 17 km (11 miles) above the Martian topographic datum: it is therefore over one and a half times the height of Mount Everest and its base stretches for a distance equivalent to that separating New York and Washington. Two other volcanoes form part of the Tharsis Montes system and these extend along a line running in a northeast to southwest direction. The Elysium Mons volcano is situated at a latitude of approximately 25°N, in Utopia Planitia. It has an interesting feature: many large canals run from the Elysium complex for hundreds of kilometers in a northwesterly direction. It is probable that the heat of the volcano melted the ice and the water scoured out these canals. Both Elysium Mons and the volcanoes of the Tharsis Montes are shield volcanoes, like those on the island of Hawaii, which were formed almost exclusively by very fluid lava. Several hundred kilometers to the northwest, at a latitude of approximately 20°, is the gigantic Olympus Mons, which is almost perfectly circular, with a diameter of 600 km (375 miles) and a height of 21 km (13 miles) above the datum; this is a colossal mountain, three times as high as Mount Everest. The caldera at its summit is 90 km (55 miles)

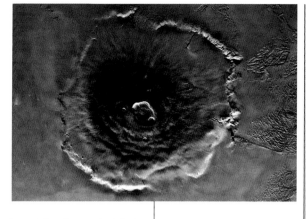

Above: Another picture of the caldera of the Olympus Mons volcano taken from a vertical position directly above it.

wide. Apart from their larger size, Martian volcanoes are very similar to those on Earth. This difference in size is probably due in part to the absence of plate tectonics. On Earth the movements of these plates slowly move the volcano away from the region in which the magma lies. On Mars the volcano remains stable, *in situ* over the source of the magma and continues to grow as more and more magma reaches the surface.

Wind

Other features typical of the Martian surface are caused by the action of wind which, as is the case in the terrestrial deserts, cause dunes, with orientations indicative of the direction of the prevailing winds. These are anything from 500 meters (500 yards–1,500 ft) to 1 km (1,000 yards–3,000 ft) wide and stretch for as far as 50 km (30 miles). Similar structures also occur on Earth, in the driest desert regions.

Above: The Twin Peaks are low hills to the southeast of the landing site of the Mars Pathfinder. They were sighted on June 4, 1997 and were compared with the pictures taken by the Viking Orbiter 20 years earlier. Their "peaks" are approximately 30 meters (100 ft) high and sited at difference distances from the observation point: one is 860 meters (940 yds) away, the other approximately 1 kilometer (1,000 yds).

Canals and gullies

These canals, which look exactly like dry river beds, are among the most intriguing features of the Martian surface. If, in fact, they were rivers long ago, the atmosphere and climate on Mars must once have been very different from what it is now, since under present conditions, as we have already seen, water is not stable in its liquid state.

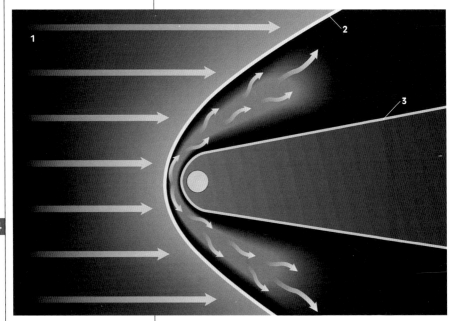

Above: Phobos 2 *did not confirm the existence of a magnetic field. The terrestrial magnetopause is therefore indicated on Mars by the deliberately ambiguous name of "planetopause."*
1. Solar wind
2. Shock wave
3. Planetopause

These Martian channels or "canals" have been compared with the largest terrestrial rivers and are gigantic in scale. For example, one of the largest is Kasei Vallis which at certain points is 100 km (60 miles) wide. Nearly all of them are as wide at their beginning as they are along the entire length of their courses; they contain areas of higher ground like little islands, their banks are well-defined and their beds show distinct signs of having been subjected to a form of scouring or "washing" along their entire lengths. When these canals were observed for the first time, the hypothesis that such extensive formations could have been caused by floods met with great skepticism: their close resemblance to terrestrial formations caused by catastrophic floods has, however, led to this explanation being given more serious consideration. It is probable that these canals were formed as a result of water being expelled from underground sources that were subjected to great pressures. Another type of Martian river bed takes the form of winding gullies which can vary in length from well under one hundred kilometers to hundreds of kilometers and these occur in the most ancient regions where there are many craters; they resemble the water courses of the torrents and high mountain streams of Earth which collect the melt-water from glaciers and rainwater. It may be that these Martian gullies have a similar origin: this would demonstrate that during the period when most of the craters were formed, that is to say 3.8 billion or more years ago, Mars had an atmosphere that was more dense, possessing a more efficient greenhouse effect and, in consequence, a less extreme climate in which water in a liquid state was stable. In June 2000, recent geological structures were identified by NASA's *Mars Global Surveyor*, probably formed by the escape of water from near-surface sources.

Canyons

The great system of canyons called Valles Marineris covers one-fifth of the circumference of Mars, at a latitude of approximately 10° south and stretches almost parallel with the equator for 4,000 km (2,500 miles). This system of canyons starts in the east at the volcanic complex of Tharsis Montes and ends in uneven terrain situated between the Chryse Planitia plain and the Margaritifer Sinus. At its deepest it is 7 km (4.3 miles) deep, the individual canyons can be as wide as 200 km (120 miles) and, in the central zone where there are three parallel, inter-linked fissures, it attains a width of 700 km (450 miles). In order to give a realistic idea of the dimensions of this canyon system, it suffices to say that the entire state of Arizona would fit easily inside the Vallis Marineris. When compared with the Grand Canyon, Valles Marineris is four times deeper, six times wider and ten times longer. Judging by the appearance of this colossal system of canyons, it would seem that it must have originated through the formation of faults and possibly also by erosion phenomena: perhaps further proof that water was abundant in far-off times. There are also signs that the canyons could have contained lakes.

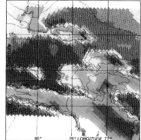

Below: Two images of Valles Marineris taken by Phobos 2. In the one on the left, absorption by the carbon dioxide increases with altitude. The higher zones are shown in red, the lower ones in dark blue. In the image on the right, the dark-blue regions indicate the presence of minerals containing compounds formed by the OH process of hydroxylation. There is no trace of water.

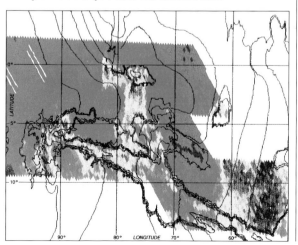

Internal Structure

Little is known about the interior of Mars. Its mean density of 3.95 g/cm^3, is slightly higher than that of the Moon, and indicates a composition more similar to that of the Moon than Earth. In order to know whether Mars contains a ferrous core and whether it is solid or liquid, it would be necessary to place seismographs where they could record Martian earthquakes. This experiment will be attempted by future space missions to the red planet: it is probable that the enormous mass of the Tharsis complex and of Olympus Mons would give rise to seismic activity. Another important difference between the internal structure of Mars and Earth concerns the thickness of the crust. The absence of tectonic plates suggests that the Martian crust must be far thicker than the terrestrial crust. Given the low gravity of Mars, it has been estimated that its crust would have to be approximately 200 km (120 miles) thick in order to provide sufficient pressure to cause the expulsion of magma from the volcanoes. The presence of a ferrous core could be proved by the presence of a magnetic field. However, data from the US space probe *Mariner 4* suggested that the Martian magnetic field was very weak, less than three ten-thousandths of the terrestrial magnetic field. Furthermore, it is not clear whether this field is generated in the interior of Mars, or whether it is imported by the solar wind. The Russian probe *Phobos*, which was equipped with a magnetometer was also unable to provide an answer to this question.

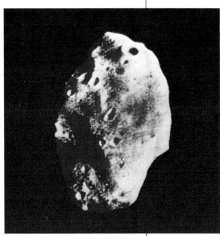

Below: Two pictures of the satellite Phobos which has a very irregular shape and is pockmarked by craters.

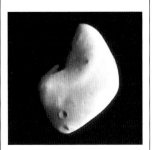

Above: A picture of the other satellite of Mars, Deimos, which is smaller than Phobos and very irregular in shape.

Is there life on Mars?

Of all the bodies of the Solar System, Mars is the most likely planet other than Earth capable of supporting living organisms, albeit primitive ones. Although there is not general agreement among biologists over what is meant here by life, according to the most widely-accepted criteria, terrestrial life is defined by the capacity of an organism to reproduce itself and by its ability to evolve through natural selection. The two *Viking* probes carried instruments which could detect life forms, if present, but results were negative or ambiguous and organic molecules were not shown to be present. According to some scientists the polar regions could be the most likely to support life. Only when Mars can be explored in the future with robots or by astronauts might it be possible to give a definitive answer. In 1996 NASA, while studying the ALH 84001 meteorite of Martian origin found in Antarctica in 1984, announced that fossilized micro-organisms might be present in it. Various scientists, however, found these results unconvincing.

Satellites

Mars has two small satellites, Phobos (from the Greek, meaning fear) and Deimos (also Greek, meaning terror), discovered by Asaph Hall in 1877; these were the names given to the two attendants of the god of war, Mars. They are irregular in shape and consequently when defining their dimensions, references are to the mean diameter: in the case of Phobos, this is 22 km (13.6 miles) and that of Deimos is 14 km (8.7 miles). Phobos rotates around Mars in 7 hours 39 minutes along an orbit that is virtually circular with a radius of 9,720 km (6,040 miles). Thus, during the course of a Martian day, Phobos completes three revolutions. Deimos is more distant and has a period of revolution of 30 hours 17 minutes, describing a near-circular orbit with a 23,400-km (14,540-mile) radius. The orbits of these two satellites are inclined by less than 2° in relation to the Martian equator. Both satellites, similar to the relationship between the Moon and the Earth, always have the same face turned towards the planet. Both Phobos and Deimos are very dark in color and have an average density of less than 2.5 g/cm3, similar to that of the meteorites known as carbonaceous chondrites. Their surface is pockmarked with craters. The crater called Stickney (which was the maiden name of Hall's wife) on Phobos is particuarly noteworthy as it measures 10 km (6 miles) across, almost half the satellite's diameter. From a point situated at the antipodes of Stickney, a series of grooves radiate outwards, which could be explained by fractures caused by the meteorite

Satellites of Mars

Satellites	Phobos	Deimos
Mean distance from Mars (km)	9.270	23.400
Distance in R_M	2.73	6.90
Orbital period (days)	0.32	1.26
Orbital inclination (°)	1.1	1.8
Orbital eccentricity	0.02	0.003
Radius (km)	10 x 11 x 14	5 x 6 x 8

Above: Four color images of Phobos taken by the Soviet probe Phobos 2, *showing that the interior of Phobos must consist of an agglomeration of different materials.*

Below: A section of the panoramic view taken by the Mars Pathfinder *over three days. The probe's image acquisition system has a definition level of 2 millimeters (0.1 inches) at a distance of 2 meters (7 feet).*

Right: The Sojourner *meets an obstacle: the bulk of Yogi. In the background, approximately 1 km (1,000 yds) away, are the Twin Peaks while below, in the foreground, are the deflated air bags and the rover's descent ramp.*

impact from which Phobos narrowly escaped destruction. Deimos has been observed from a distance of less than 30 km (20 miles) by the *Viking 2* orbiting probe: the pictures it obtained show a surface peppered with craters, most of which are shallow and covered in a "dusty" regolith. The escape velocities of Phobos and Deimos are, respectively, 15 m/s (50 feet/s) and 10 m/s (33 feet/s). The fine powdery regolith that covers both of these satellites must possess a higher velocity than the escape velocity, because what remains is thought to form a very thin layer less than a millimeter deep. It is not clear, however, why the craters on Deimos have a thicker covering of regolith than those on Phobos. As to the origins of these two Martian satellites, some astronomers think that these are asteroids trapped by Mars, but the fact that both of them orbit almost in the plane of the Martian equator leads others to believe that they probably share their planet's origin.

Right: A picture taken by the Mars Global Surveyor. *The valleys and plateaus in the Gorgonum Chaos region are clearly visible. It is believed that the canyons were formed as a result of water penetration from a layer sited near the upper edges of the walls of each valley. The presence of numerous depressions suggests that the layer situated only a few hundred meters below the surface is very permeable, with considerable water-storage capacity.*

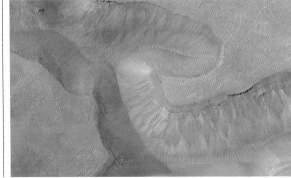

IN SEARCH OF MARTIANS

After the US Mariner and Soviet Mars probes, NASA launched Viking 1 and 2 which landed on the Red Planet in 1976. These were real space robots, equipped with chemical laboratories, complex technical equipment and an extendable arm with which to gather soil samples for analysis. The landing module also carried a seismograph and a television camera which could compose an entire image in black and white, in color, and also in infrared using a digital technique, in 20 minutes. The two probes landed at two different sites on the surface and their operational life continued until the beginning of the 1980s, long after their projected life span of 90 days; between them, they gathered 4,500 images, but the biochemical data proved disappointing for the exobiologists. The tests which appeared to show biological activity were contradicted by later analyses, according to which only inorganic reactions due to the particular composition of the regolith took place on Mars. After the failure of the Mars Observer in 1993, NASA decided that the next missions were to use an orbiting probe, and a roving vehicle to explore the surface. The Mars Global Surveyor, launched in November 1996, carried out detailed mapping and, of course, surveying of many aspects of the planet; it discovered strong indications of the presence of water on the surface of the planet. The Pathfinder probe or lander, which was a meteorological and environmental station with the Sojourner mini-rover on board, bounced down onto the ground on Mars, surrounded by protective airbags, in July 1997. Pathfinder sent 2.6 million bytes of data back to Earth, among which were 16,000 images and Sojourner sent back 550. This rover was 48 centimeters (18 inches) long, traveled at a speed of 1cm/s (0.4 inch/s) and carried out 15 chemical analyses of the rocks encountered in the landing area, Ares Vallis.

Among the large amount of data gathered, it was concluded that the surface had been worn away and carved out during far-off eras by raging torrents of fast-flowing water. After the failure of two Martian probes in 1999, NASA resumed exploration in April 2001, carrying out a mineralogical reconnaissance with its orbiting Mars Odyssey probe.

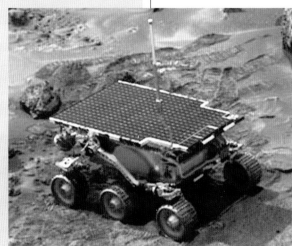

Below: A photograph of the Rover Sojourner, taken by Pathfinder. This picture makes it possible to differentiate between the colors of the Martian landscape, ranging from the dark rocks to the bright red powder covering the darker ground. Below the edge of the solar panel, in shadow, it is just possible to make out the television cameras.

THE OUTER PLANETS

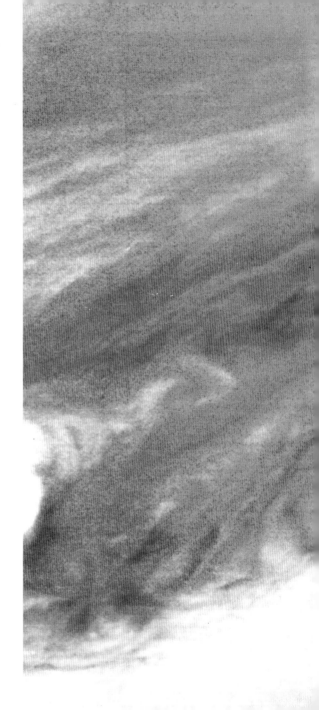

JUPITER ♃

Above: Jupiter as viewed from the Voyager 1 probe at a distance of 28.4 million kilometers (17.6 million miles). The easiest features to observe are the dark bands and the pale zones parallel to the equator, which are simply belts of differently-colored clouds. The small oval or circular spots indicate areas of turbulence.

Physical characteristics

Jupiter has a mass 318 times greater than the Earth's and a diameter that is 11 times larger. Its volume would therefore be capable of containing 1300 planets the size of the Earth. Jupiter's mass is 70% of the sum total of mass of all the other planets in our Solar System. A mass only slightly larger would have led to the planet contracting inwards on itself through the effect of its gravitational field and, were its mass considerably larger it would have formed a star. Jupiter's actual mass and distance from the Sun means that this planet is able to retain hydrogen in its atmosphere, together with helium. It is believed that the chemical composition of the Jovian atmosphere is the same as the primordial matter from which the Solar System was born, approximately 4.5 billion years ago. At a mean distance of 5.203 astronomical units from the Sun, equivalent to 778,358,238 km (483,617,809 miles), Jupiter takes nearly 12 years to complete an orbit. The closest this giant comes to the Earth is 591,000,000 km (367,247,000 miles) and the furthest it travels away from us is 967,000,000 km (600,894,000 miles). During the period 1980-1992 Jupiter was in opposition 11 times with a magnitude of −2 and, in 1987, as much as −2.5. From 1993 to 2000 it was possible to observe it in opposition a further eight times with a magnitude of −2.5 during the two years 1998-99.

Rotation

Jupiter rotates faster than all the other planets in the Solar System. A relatively small telescope is adequate to discern bands of clouds around the planet on which can be observed semi-permanent marks or spots. The movement of these spots around the planet's axis indicate that its rotational periods are variable and are shorter around the equator. From this astronomers have inferred that Jupiter does not rotate like a solid body, but behaves more like the Sun. Unlike the Sun, however, with its rotation period that gradually changes according to latitude, Jupiter's rotation period varies from zone to zone and although the variations from zone to zone are reminiscent of the terrestrial atmosphere, they are more complex.

The period of rotation observed within 10° from the equator is approximately 9 hours 50 minutes, similar to the poles. This high rotational velocity, combined with Jupiter's low density, equal to 1.3 times that of water, causes a significant flattening of the planet, which has an equatorial diameter of approximately 142,000 km (88,000 miles), while that of its poles measures 134,000 km (83,000 miles). Jupiter possesses a very strong magnetic field; this is a dipolar field like that of the Earth, but with reversed polarity and consequently, were a terrestrial compass used on Jupiter, it would point to the south instead of the north. Jupiter is not a single body, but a system surrounded by two rings and at least 16 satellites, amongst which are the four great Galilean moons. This is

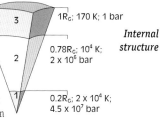

Internal structure

$1R_G$; 170 K; 1 bar
$0.78R_G$; 10^4 K; 2×10^6 bar
$0.2R_G$; 2×10^4 K; 4.5×10^7 bar

1—Core of silicates, minerals and ice
2—Metallic hydrogen,
3—Molecular hydrogen

Atmosphere

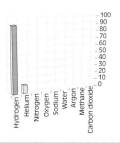

Characteristics of Jupiter

Mean distance from Sun (AU)	5.20	Mass (g)	1.9×10^{30}
Mean distance from Sun (10^6 km)	778.33	Mass (Earth = 1)	317.938
Orbital period (days)	43332.71	Equatorial radius (km)	17,492
Mean orbital velocity (km/s)	13.07	Equatorial radius (Earth = 1)	11.209
Orbital eccentricity	0.0483	Mean density (g/cm³)	1.33
Apparent mean diameter of Sun	6' 09"	Mean density (Earth = 1)	0.24
Inclination of orbit to ecliptic (°)	1.308	Volume (Earth = 1)	1,319.6
Number of satellites	16	Ellipticity*	0.0649

Equatorial surface gravity (m/s²)	22.88
Equatorial surface gravity (Earth = 1)	2.34
Equatorial escape velocity (km/s)	59.6
Sidereal rotation period at equator	9 h 50 min 28 s
Inclination of equator to orbit (°)	3.12

*Ellipticity is (Re—Rp)/Re, where Re and Rp are the planet's equatorial and polar radii, respectively.

Above: A comparison of the size and rotational axial inclination of Jupiter and the Earth.

Below: Wind speeds in the various zones and at different latitudes on Jupiter, as observed by Voyager 1 on February 26, 1979 when the probe was 7.8 million km (4.8 million miles) from the planet. The maximum velocity occurs in the equatorial zone, where it reaches 150 meters (500 feet) per second (540 km/h or 340 miles/h).

the reason why Jupiter is described as a solar system in miniature, although the same could also be said of the other giant planets Saturn, Uranus and Neptune.

Surface

The surface of Jupiter or, more accurately, its upper atmosphere, is characterized by the presence of variously-colored clouds, forming bands running parallel to the equator, with the exception of the polar regions. The paler bands, formed of ascending, hotter gases and referred to as "zones," alternate with darker, reddish bands of cold, descending gases, known as "belts". The atmosphere of a terrestrial planet accounts for a negligible contribution to their mass but in the case of Jupiter and the other giant planets, their atmospheres are much more extensive.

Jupiter's dense clouds form in an atmosphere that has been calculated to consist of 88% molecular hydrogen, 11% helium and the remaining 1% of methane, ammonia, water, carbon dioxide and other substances: an atmosphere similar to that which the Earth possessed during the first 100 million years of its existence. When these residual elements, and sulfur in particular, combine with the hydrogen atoms, they "dye" the clouds with colors shading from red to deep orange, from brown to yellow, from green to dark blue and all the shades in between.

The internal heat of the planet generates powerful convection currents which continually stir up and shift around the clouds and, because

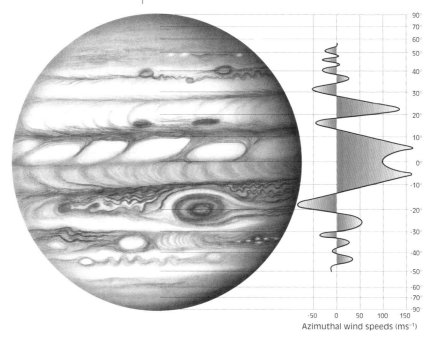

Azimuthal wind speeds (ms^{-1})

of Jupiter's rapid rotation, they are stretched out into belts and zones around the planet. The clouds that start off as red or brown or dark blue, having formed at a certain depth and temperature therefore undergo further chemical transformations, growing paler towards white or changing

Left: A color image of Jupiter obtained with the Hubble Space Telescope's Wide Field and Planetary Camera.

Below: A detail of the Hubble Space Telescope's image of Jupiter which shows the planet's southeast quadrant. Of particular interest: the dark, oval ring, which, like the Great Red Spot and other similar structures, is probably a cyclonic formation and its coloring is due to the presence of sulfur, phosphorus and carbon in the form of ice crystals.

to other colors at differing latitudes and altitudes. The northern tropical band is usually more easily discernible than the symmetrical southern band. Occasional gaps in the cloud blanket in regions of low pressure, such as the dark oval spots, sometimes enable observers to glimpse, or to deduce from spectroscopic data, what the deeper atmospheric layers look like. Were it possible to observe Jupiter from a position located where the stratosphere meets the troposphere, this would confirm that all meteorological activity takes place under the Jovian stratosphere, as is the case on Earth.

Clouds and winds

Approaching the level of the higher clouds, where the temperature hovers around 130–150°C (200–240°F) below zero and pressure is approximately 50 millibars, or five-hundredths of Earth's atmosphere; the color of the sky is probably blue due to the scattering of sunlight. Then, however, between 30 and 60 km (about 20 and 40 miles) below the altitude where the stratosphere and troposphere meet, a first layer of red and white clouds full of ammonia snow flakes occurs, followed lower down by a layer of brown clouds of crystals of ammonium sulfide. Finally, approximately 90 km (55 miles) from the ionosphere, there is a third layer of bluish clouds formed of water droplets and flakes of frozen snow. It is possible that these layers do not only contain inorganic substances, organic molecules could also be present. Because of the planet's rapid rotation, strong trade winds and jet streams blow between the belts and

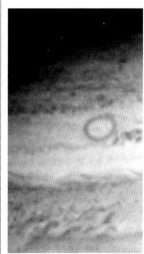

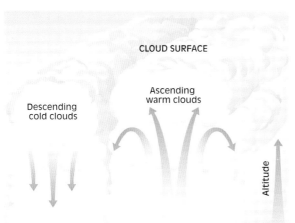

Above: A diagrammatic illustration showing how the cloud belts of Jupiter are formed. The warm, rising clouds produce the pale bands; the cold, falling clouds turn to dark bands. The direction is then reversed, because the warm rising clouds cool and the cold falling clouds come into contact with the lower layers and heat up.

Below: A detail of the Great Red Spot observed by Voyager 1. Towards the bottom of this image is an almost perfectly oval white spot.

the zones, surrounding the entire planet with cyclonic and anticyclonic phenomena associated with low and high pressures respectively. The former are typical of the zones, the latter of the belts. At the higher polar latitudes, this regular pattern of high and low pressure is interrupted, causing the formation of chaotic hurricanes. Unlike Earth, Jupiter's polar and tropical temperatures are virtually the same.

The Great Red Spot

Since the 1960s it has been known that Jupiter radiates two-and-a-half times the energy that it receives from the Sun. It is possible that this excess comes from the energy produced during the formation of the planet, which is still being emitted. Another hypothesis, albeit less plausible, could be that of a slow and continuous contraction of the planet which could thus disperse heat outwards, to the exterior. Heat generation, whatever its origin, not only gives rise to convection phenomena, to the transfer of colored materials to the highest clouds, and to atmospheric whirlwinds and cyclones, but also to hurricanes such as the Great Red Spot which whirls over the surrounding clouds and has persisted for at least three centuries. Its color varies from brick red to greenish pink and its diameter exceeds that of the Earth.

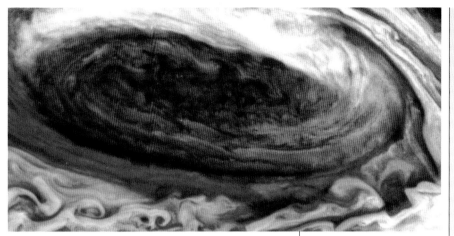

Above: A further detail of the Great Red Spot. The presence of a vortex is very clearly shown.

The Great Red Spot moves westwards and eastwards at a speed of approximately one meter (3.3 feet) a second in relation to the other structures, but it never moves northwards or southwards: it has therefore already circled the planet many times. Other, similar oval spots exist, while they are not as persistent, they are always a sign of cyclonic storms (in which case they are darker in appearance) or of anticyclonic storms (when they are lighter). They are generated at the edges of the jet streams which move in opposite directions around Jupiter.

Properties of the Great Red Spot

The Great Red Spot rotates in an anti-clockwise direction, with a period of 12 terrestrial days and if some of its energy did not disperse into its surroundings, could last indefinitely. Fed by the great jet streams of hot gas, huge cumulo-nimbus clouds form, turning the Great Red Spot into an infernal environment, riven by electrical discharges. A few years ago, the American research chemist Cyril Ponnamperuma carried out an experiment during which he produced electrical discharges in an atmosphere of methane, ammonia and hydrogen, sealed in a glass container. The colorless mixture of gases acquired a foggy appearance and started to form a crystalline substance similar in color to that of the Great Red Spot. Chemical analysis revealed that this was an organic compound called nitril which, when combined with water produces amino-acids, the fundamental constituents of proteins in all known life forms. This leads to the conclusion that the earliest seeds of life could have originated in places such as the Great Red Spot or, even more plausibly, in similar places with lower turbulence.

Internal structure

It could well be that below Jupiter's layer of water-vapor and ammonia clouds there exists a mist of water and ammonia droplets, above a boundless ocean of hydrogen and other elements with a sludge-like consistency, covering the entire planet, over an area 114 times that of the Earth. A ship that takes one and a half months to circumnavigate the Earth at a speed of 20 knots would take nearly a year and a half to sail around Jupiter.

On this ocean of liquid molecular hydrogen almost total darkness reigns, fitfully illuminated by the lightning flashes unleashed by the lowest clouds. If it were feasible to explore this ocean in a bathysphere to avoid being squashed by the pressure, we could imagine a descent of approximately 25,000 km (15,500 miles), until the point was reached that the pressure was in excess of 2 million atmospheres, when the liquid molecular hydrogen changes into liquid metallic hydrogen, a mixture of protons and electrons, circulating freely, creating electrical currents: at very high pressures hydrogen does, in fact, behave like a metal. Descending even deeper, for another 37,000 km (23,000 miles), the explorer would reach a rocky core with a radius of approximately 12,000 km (7,500 miles), probably composed of iron silicates. Here the temperature reaches 30,000°C (55,000°F), while in the transition zone between the liquid molecular hydrogen and liquid metallic hydrogen it fluctuates around 11,000°C (20,000°F).

The interior of Jupiter therefore also has a three-layered structure. A rocky core, larger than the Earth's, is surrounded by a layer of liquid

Right: Jupiter possesses a strong magnetic field, approximately 12 times greater than that of the terrestrial field (4.2 gauss as opposed to Earth's 0.35). It has an axial inclination of 11° in relation to its rotational axis. The presence of liquid metallic hydrogen in the interior of Jupiter is an important factor in accounting for the presence of a magnetic field. This fluid is, in fact, an electrical conductor and Jupiter's rapid rotation can generate or provide favorable conditions for the development of electrical currents which create magnetic fields.

metallic hydrogen and by another layer of liquid molecular hydrogen. Above this extends an atmosphere that, up to the altitude of its visible surface, is approximately 1,000 km (600 miles) high. The electrical currents that are present in the layer of liquid metallic hydrogen, together with Jupiter's rapid rotation, produce its strong magnetic field through what can best be described as a dynamo effect.

The magnetic field

Jupiter's magnetic field forms an angle of nearly 10° relative to the planet's rotational axis and is approximately 10 times stronger than the

Earth's field. Like the terrestrial field it is also dipolar, although other fields have been detected in giant planet's vicinity which are quadrupolar and octupolar.

The magnetic field envelopes the planet in a gigantic magnetosphere, large enough to extend beyond the orbits of many of its moons and, in common with all planetary magnetospheres, this has a shock wave that faces the solar wind. If the latter varies in strength, the magnetosphere shrinks further back, away from the Sun, or extends further out, towards the Sun, a variation in distance of between 25 and 50 planetary diameters, or between 3 and 6 million kilometers (1,800,000 and 3,600,000 miles): well beyond its satellite Callisto and as far as the near side of its satellite Leda.

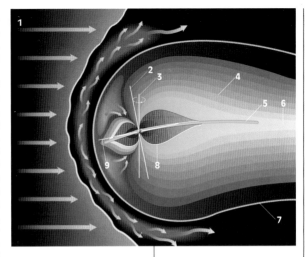

Above: Jupiter's magnetosphere
1. Solar wind
2. Magnetic axis
3. Rotational axis
4. Force lines of the magnetic field
5. Neutral zone
6. Magnetotail
7. Magnetopause
8. Zone of intense radiation
8. Magnetic equator

Structure of the magnetosphere

The action of the solar wind on the magnetic field of Jupiter creates the planet's magnetosphere. The shape of Jupiter's magnetosphere is similar to that of the Earth and consists of a magnetopause, a disk of plasma and a magnetic tail which stretches out beyond Saturn. There are, however, significant differences when compared with the terrestrial magnetosphere. Since Jupiter rotates much faster than our planet, it drags part of the magnetospheric plasma around with it, concentrating this into a disk, or torus, of electrical current in the magnetic equatorial plane. If this enormous disk were visible, it would appear larger than the Sun or the Moon. Since the disk does not lie exactly on the equatorial plane, because of the inclination of the magnetic axis in relation to the rotational axis, it seems to move around like "the enormous brim of a hat

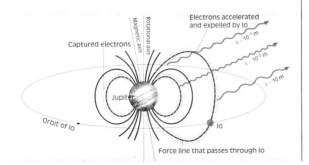

Left: The presence of a Jovian magnetic field was detected for the first time through observations of its radio emissions. The high-energy particles which abound in the vicinity of Jupiter are trapped by its magnetic field and produce radio-electric emissions of varying wavelengths. The presence inside Jupiter's ionosphere of the satellite Io and the ring of ionized gas that surrounds it, produces other, higher-energy radio waves, having wavelengths between 0.1 and 1 meter, and microwaves as shown in the diagram.

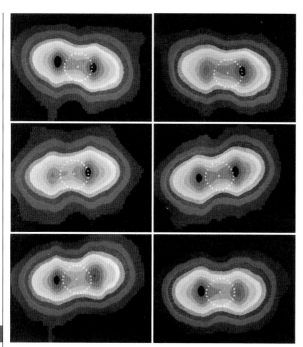

Above: The distribution in intensity of radiation with wavelengths of 20 cm, observed by the Westerbrook radio-telescope in Holland is shown in conventional colors. The ring of dots stands for the outline of the planet. The variable tilt of the image is due to the fact that the magnetic axis is inclined by 11° in relation to the rotational axis.

Below: Jupiter also emits X-rays. Observations carried out by the Einstein satellite show that X-ray emissions emanate from the high-latitude regions of the planet. They imply the existence of polar aurorae similar to those seen on Earth.

placed somewhat crookedly on the head of a pirouetting skater," as Roman Smoluchowski has described it.

Before the *Voyager* probe's reconnaissance it was thought that some particles of the solar wind, escaping from the tail of the magnetosphere, penetrated into the polar regions, creating aurora borealis, as happens on Earth. The reality is quite different: it was discovered that Jupiter's aurora borealis are produced as a result of the interaction between Jupiter's ionosphere and the electrons of the torus, or ring of plasma, in the orbit of Io. This satellite is, moreover, directly linked with Jupiter's magnetic poles by an electrical flux of current of a staggering 5 million amperes. X-ray emissions have also been detected in regions at high latitudes coinciding with polar aurorae; these emissions seem to be identical to those encountered on Earth. The night side of Jupiter displays these extraordinary aurorae which we can only imagine in all their phantasmagoric colors, together with their whistles, crackles, flashes and flickers, like a garish neon sign. On this side there are also fearsome lightning flashes darting through the clouds, and the blazing trails of meteorites attracted by Jupiter's powerful gravitational field. Even the well-known asteroid belt between Mars and Jupiter is disturbed by the great planet's gravitational field, which has also changed the orbit of at least fifty known comets, causing what were at one time long-period comets to become short-period comets.

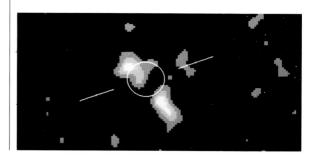

The rings of Jupiter

The rings of Jupiter are so evanescent in appearance that prior to the discoveries made by the *Voyager* probes, they had never been seen and their existence was not even suspected; after the *Voyager* missions, they have been observed from Earth, using powerful telescopes to search for them.

The well-defined outline of the outermost ring reaches a distance of 55,000 km (35,000 miles) above the planet's clouds. It is the brightest ring and is approximately 800 km (500 miles) wide; the second ring from the outside is approximately 6,000 km (4,000 miles) wide; there is a third, even fainter ring, which seems to stretch inwards as far as the planet's atmosphere. The rings of Jupiter are of a distinctive orange color and it would appear that they are no more than a few kilometers deep. The minute particles of which they are formed orbit around the gigantic planet with a period that varies from 5 to 7 hours. It is estimated that these particles must be so small in size that they are unlikely to come from any source other than volcanic eruptions on Io: the material expelled from Io must have bombarded a small satellite on the outer edge of the ring and carried dust and fragments away from it. While the fragments created the outermost, brightest ring, the dust is thought to have formed the inner and more opaque rings. The particles which form the rings of Jupiter are therefore not believed to consist of ice but of a rocky substance.

Left: The weak ring of Jupiter was observed by Voyager 2. It can be clearly seen, stretching out on both sides of the planet.

Below: The same ring in greater detail. The main structure is indicated by the dark-gray band; it has an extensive envelope in the form of a torus, which is barely perceptible and, a thin "ethereal" ring (shown in lighter gray) which extends almost as far as Amalthea's orbit, approximately 180,000 km (111,000 miles) from the center of Jupiter.

Ori. ✱　　　✱　　✱○　　　✱　　Occ.

Above: Galileo recorded his first series of observations of Jupiter, in January 1609, in his Sidereus Nuncius, *as shown above.*

Satellites

Jupiter reigns over a court of at least 16 satellites. Other satellites may be discovered by ground-based telescopes or during future space exploration missions. Their density is estimated as equivalent to three times that of water and they are thought to consist of rocky material rather than of ice. Situated on the edge of the outermost ring of Jupiter's rings, Adrastea marks its edge like a milestone and, together with Metis, plays a part in maintaining the stability and persistence of the rings.

The rings are composed of dust that falls in a continual spiral onto the planet and their life would be foreshortened were it not for the presence of Adrastea, and Metis which is sited further in, which are bombarded by micrometeorites and keep the rings continually supplied with more dust. The most obvious features of Amalthea are its irregular shape and its pitted surface. Its major axis is tilted towards Jupiter, in conformity with tidal laws. This satellite is, in fact, close to the "Roche limit:" the distance from the center of a planet within which any relatively large satellite would be torn apart by tidal forces. The temperature of Amalthea is higher than can be accounted for solely by incident radiation from the Sun and Jupiter. It is possible that this anomaly is caused by the electrical current induced by Jupiter's magnetic field. Amalthea's surface is

Jupiter's Satellites

Satellites	Mean distance from Jupiter (km)	Distance in R_G	Orbital period (days)	Orbital inclination (°)
Metis	127,960	1.79	0.3	0
Adrastea	128,980	1.8	0.3	0
Amalthea	181,300	2.54	0.5	0.4
Thebe	221,900	3.1	0.68	0.8
Io	421,600	5.9	1.77	0.04
Europa	670,900	9.38	3.55	0.47
Ganymede	1,070,000	14.97	7.16	0.19
Callisto	1,883,000	26.34	16.69	0.28
Leda	11,094,000	155.18	238.72	27
Himalia	11,480,000	160.58	250.57	28
Lysithea	11,720,000	163.93	259.22	29
Elara	11,737,000	164.17	259.65	28
Ananke	21,200,000	296.54	631	147
Carme	22,600,000	316.12	692	163
Pasiphae	23,500,000	328.71	735	147
Sinope	23,700,000	331.51	758	153

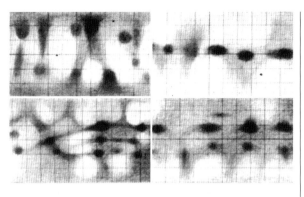

Left: A map of the four Galilean satellites taken from observations recorded by Lyot from the Pic du Midi Observatory. When compared with more recent images sent back by space probes, their general appearance shows some similarity.

Below: A scale diagram of Jupiter's satellites, showing their comparative sizes.

dark red in color, perhaps due to sulfur dust originating on Io, and it is heavily scarred by impact craters, including the enormous Pan and Gea craters, which are deeper than similar lunar craters.

When seen from Jupiter, Amalthea has an apparent diameter of 7'24", while that of the Sun has a smaller apparent diameter of 6'; thus, Amalthea causes total solar eclipses on Jupiter, as do the four largest of the Galilean satellites. Apart from its craters, Amalthea's surface is furrowed with faults and ridges many kilometers in length, attaining widths of up to 20 km (12 miles). If the satellite's mass were known, this would reveal whether it is a captured asteroid or a satellite that has formed *in*

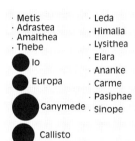

- Metis
- Adrastea
- Amalthea
- Thebe
- Io
- Europa
- Ganymede
- Callisto
- Leda
- Himalia
- Lysithea
- Elara
- Ananke
- Carme
- Pasiphae
- Sinope

Orbital eccentricity	Radius (km)	Mass (g)	Mass (Luna = 1)	Mean density (g/cm³)	Mean density (Luna = 1)
0	20				
0	12 x 10 x 8				
0	135 x 85 x 5				
0.01	55 x 50 x 45				
0	1815	8.94 x 10^{25}	1.22	3.57	1.07
0.01	1569	4.80 x 10^{25}	0.65	2.97	0.89
0	2631	1.48 x 10^{26}	2.01	1.94	0.58
0.01	2400	1.08 x 10^{26}	1.47	1.86	0.56
0.15	8				
0.16	90				
0.11	20				
0.21	40				
0.17	15				
0.21	22				
0.38	35				
0.28	20				

situ. Virtually nothing is known of Thebe, apart from its orbital and rotational periods (Thebe moves between Amalthea and Io) and its size, which is considerably larger than Metis and Adrastea. After the four great Galilean satellites which are described below, come eight outer satellites: Leda, Himalia, Lysithea, Elara, Ananke, Carme, Pasiphae and Sinope. Names ending in the letter "e" denote those satellites with pro-

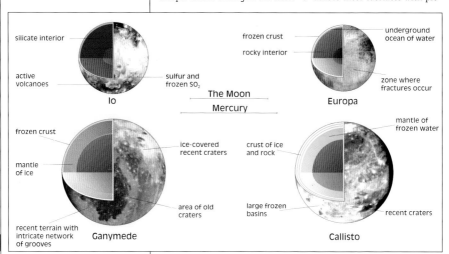

grade or direct motion, those ending with the letter "a," denote retrograde motion.

The first four of these satellites are separated from Callisto and from the other, inner satellites by a vast gap. Their orbits wind in and out of each other and their orbital planes are inclined by approximately 28° on the equatorial plane of the planet. Ananke, Carme, Pasiphae and Sinope are at a distance from Jupiter that makes them particularly vulnerable to disturbances, so much so that their respective orbits are easily altered. An example is provided by Pasiphae, which was lost after it had been discovered in 1908, and was re-discovered after 14 years, in 1922, only to disappear once more in 1938, and yet again between 1941 and 1955. As to the origin of these satellites, there are two theories: they could be asteroids captured by Jupiter, or could have resulted from a collision between asteroids or between satellites. Both these hypotheses are, however, problematic and, in the absence of further observations, this is a question that has yet to be answered.

The four Galilean satellites

Generally speaking, the evolution of the celestial bodies is determined by their mass. A body of small mass, lacking sufficient sources of internal energy, does not evolve and soon cools down. There are, however, many cases in which mass is not the only thing that counts, another fac-

Above: The four Galilean satellites: Io, Europa, Ganymede and Callisto. The first two are of virtually the same size and the same density as the Moon and are mainly composed of rocky material. The other two are approximately the size of Mercury, but less dense, and appear to be covered by a thick layer of ice and water.

Below: An image of Amalthea, one of Jupiter's smallest satellites, orbiting within Io's orbit.

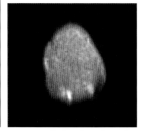

tor being the presence of other bodies in greater or lesser proximity which induce energy, either by radiation or through the action of tidal forces, electrical fields, collisions and so on. An example is provided by Io which, along with Europa, Ganymede and Callisto form the quartet known as the great Galilean Satellites.

The orbital period of Europa is twice that of Io, while Ganymede has a period double that of Europa. This is no coincidence: it is due to a phenomenon known as "resononance" or complex interactions between tidal forces which have been established between the reciprocal masses and that of Jupiter. Each of these satellites is in synchronous rotation, keeping one face turned toward the planet and all of them, with the exception of Europa, are larger than the Moon. Like the Moon, they are without atmospheres, with the possible exception of Io which appears to possess one, although very tenuous. Their densities range from 3.5 g/cm3 for Io, to 1.81 g/cm3 for Callisto, with a noticeable diminution proportional to its distance from Jupiter. Since Io and Europa have densities similar to that of the Moon, scientists speculate that they are composed of the same rocky material, unlike Ganymede and Callisto which are probably made of ice.

Io

Of all the satellites in our Solar System, Io is certainly the most spectacular in volcanic activity. Even when the *Voyager 1* probe was still a long way off from this satellite, it was obvious that its surface was completely different from that of the other Galilean satellites. Io is a bright orange color and has no craters; it therefore has all the appearance of a young surface that formed after the period of meteorite bombardment which hammered the Solar System during its formation. However, Io's surface is punctuated by active volcanoes. This constant volcanic activity had already been forecast by certain scientists, notably by Stanton Peale who put forward the hypothesis that this is caused by the attrition of the tidal stresses exerted by Jupiter and by the other Galilean satellites, especially by Europa. As a result of such forces, he maintained, Io's crust must lift and then fall back, producing heat which, apart from liquefying Io's core, would also escape through the exterior of the satellite by means of volcanic eruptions. Nine volcanoes have, in fact, been photographed, with plumes that rose as high as 280 km (175 miles).

Below: A close-up image of Io, taken by Voyager 1 on March 4, 1979, from a distance of approximately 862,000 km (536,000 miles), just over double the distance between the Moon and the Earth. The Prometheus volcano is visible in the center of the image. Io is exceptional in that its surface is peppered with active volcanoes (the roughly circular black spots) and has regions ranging in color, from white to yellow to deep orange, caused by sulfur deposits, varying in temperature, which have erupted from the volcanoes.

The surface of Io is also covered with extensive lava flows; hot zones have been identified with a temperature of 17°C (63°F), while in the surrounding environment temperatures of 176°C (280°F) below zero have been recorded. Temperatures in the calderas of Io's volcanoes reach values in excess of 500°C (900°F). Sulfur is omnipresent, forming a powdery blanket over a silicate surface not unlike terrestrial volcanic regions. Io's volcanoes emit large quantities of sulfurous material, but approximately 100 kg (200 lb) is dispersed every second into space. This creates an insupportable atmosphere, while a blizzard of yellow, bright orange and bluish-white flakes of anhydrous sulfate falls to the ground. It is calculated that deposits of this material build up on Io's surface at the rate of approximately 0.1 cm each year. Io is also surrounded by a great cloud of sodium, calcium and hydrogen, undoubtedly of volcanic origin; in addition to this *Voyager* identified a large toroidal ring of ionized particles near its orbit. This torus of ionized sulfur extends for 5–6 Jovian radii and is confined to Jupiter's magnetic equatorial plane, inclined at an angle to the planet's equatorial plane. This ring consists of atoms of sulfur and oxygen ionized by ultraviolet solar radiation and by electrons freed of these elements, which give rise to an electrically neutral plasma.

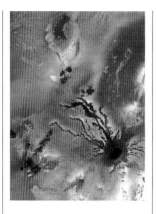

Above: An image of a volcanic crater on Io.

Below: Europa as seen from Voyager 2. Unlike Io's surface which is covered with active volcanoes and has areas of color varying from black to white to yellow and orange, Europa's coloring is fairly uniform and its surface is covered with a tangle of thin cracks. There are virtually no impact craters. The surface is flat, with no mountains.

Europa

Whitish, highly-reflective, with no volcanoes nor obvious craters, the surface of Europa looks like an Arctic ocean. Its surface is known to be covered with ice, with shallow grooved and furrowed lines similar to the "canals" of Mars; also present on Europa's surface are ridges and dark markings of uneven shape. Europa is a smooth satellite, rather like a billiard ball, making it difficult for a cartographer to map it. It is believed that Europa is covered with an ice crust and below this lies a slushy layer of water and ice and a fairly large rocky core, with a radius measuring 1,400 km (900 miles). The absence of large craters (only three have been glimpsed, none wider than 20 km/12 miles) would seem to indicate that Europa's surface dates back to an era more recent than 4 or 4.5 billion years ago, when the planets were already formed and had been immediately subjected to an intense bombardment by planetoids. Europa must have a much more recent origin. Perhaps the layer of ice slid into the oldest and deepest impact craters, just as a glacier would do. Current opinion favors the theory that the satellite was formed, like Ganymede and Callisto, from a mixture of ice and rock dust in the cold primordial cloud which orbited around Jupiter, but that it was subjected to a greater amount of heat, probably through the same tidal stresses that liquefied

Io's interior and produced its volcanoes. Whatever the true case may be, the heat that reached Europa was enough to liquefy its core, if only partially, enabling a watery "lava" to erupt and cover the surface, leveling this out and smoothing away any unevenness before changing into ice.

Ganymede

This is the largest of Jupiter's satellites. Its crust of ice is estimated to be approximately 100 km (60 miles) thick, under which there seems to be a convective layer of water or semi-liquid ice, varying in thickness between 1,800 km and 2,200 km (1,100–1,400 miles). The surface is a gray-blue color and it has a generally dirty appearance, except where

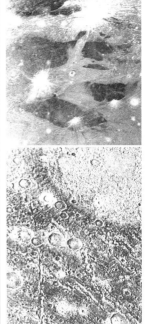

meteorite impacts have exposed the underlying, somewhat paler layers of ice. The landscape on Ganymede is scattered with craters varying from the ancient to more recent. Its surface certainly looks far older than that of Io and Europa. Other features typical of Ganymede are the whitish polar caps, which are apparently thin deposits of water hoar frost, stretching as far as latitudes of 40° north and south. The poles are the coldest regions because they receive less direct sunlight and it is possible that the water vapor that has escaped from the fractures around the planet migrates and condenses in these regions in the form of hoar frost.

Callisto

This satellite is the least dense of the Galilean satellites and probably one of the least dense satellites measured to date. It has the lowest albedo, meaning that it is the least reflective. It has numerous craters although these are shallow, and is more heavily-cratered than any other body in our Solar System. Current thinking is that Callisto has not undergone substantial changes for billions of years, which means that it is the satellite with the oldest surface in the Solar System. Its surface would appear to be composed of a crust of dirty ice and rocks approximately 300 km (180 miles) thick, under which is thought to be a mantle

Left: Ganymede as seen by Voyager 2. This image shows the face that is turned away from Jupiter. The most noticeable feature here is a large circular region, called Galileo Regio, all that remains of Ganymede's ancient crust. Above, top: This detail from the Voyager image of Ganymede shows dark regions, islands of the ancient crust of this satellite, while the more-recently formed areas of terrain are paler in color.

Above: A detail of the grooves on Ganymede's surface. Numerous mountain ridges, approximately 1,000 meters (3,300 feet) high run in parallel formation, separated by approximately 15 kilometers (10 miles). The smallest details visible on this satellite's uneven surface measure 3 km (about 2 miles). The entire area shown is the same size as Pennsylvania.

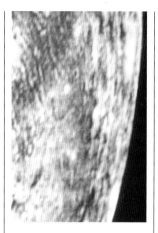

of water and ice approximately 1,000 km (600 miles) thick, and under this in turn lies a rocky core with a 1,200-km (750-mile) radius. There are two great basins on Callisto, Valhalla and Asgard. The former is surrounded by a series of concentric rings which enclose a lighter, circular region 600 km (about 400 miles) in diameter; the second, Asgard, is similar but less extensive. Callisto's surface temperature varies from 18°C (0°F) below zero during the day, to 193°C (315°F) below zero during the night.

Above: Callisto, observed from a distance of 1.2 million kilometers (750,000 miles) by Voyager 1.

Right: Jupiter in an early-twentieth-century chromolithograph.

Below: Two photographs of the planet taken between 1939 and 1956.

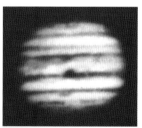

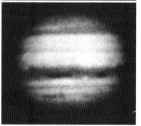

A SPOT AS LARGE AS THE EARTH

Gian Domenico Cassini was the first astronomer to pursue in-depth studies of Jupiter. He benefited greatly from being able to use Giuseppe Campani's telescopes which permitted up to 200-times enlargement. In addition to his discovery of the Great Red Spot in 1665 which enabled him to make the first estimate of 9 hours 56 minutes for the planet's rotation, he was also able to calculate a period of 9 hours 51 minutes for the structures on the equatorial belt. The difference between rotational periods at various latitudes showed that the details observed could not be part of a solid surface. Cassini, in common with Francisco Fontana before him, noted how fast the structure of the dark belts on the planet's disk could change and interpreted these as the equivalents of the clouds in the terrestrial atmosphere. The comparison between the planet's bands and its climatic zones led him to conjecture, correctly, that when observed from the exterior these must appear darker when cloud cover was thicker. He also put forward the hypothesis that the differences in rotational periods that he had observed were due to large atmospheric movements, which dragged the clouds and the structures connected with them along in their wake. The only permanent structure turned out to be what came to be known as the Great Red Spot, at times vividly visible, at others, pale. Nineteenth-century astronomers thought that it was an atmospheric structure linked to some peculiarity of the underlying surface, but this hypothesis has now been rejected.

JOURNEY TOWARDS THE GIANT OF THE SOLAR SYSTEM

In 1969 NASA gave the go-ahead for the construction of the Pioneer 10 and 11 probes which were to be sent towards Jupiter, inserted into different orbits; one of which was intended for the reconnaissance of Saturn. Various obstacles meant that this project was a stiff technological challenge: the asteroid belt had to be safely negotiated, as did the intense radiation around Jupiter; continuous and sufficient energy supplies had to be guaranteed for the journey to such a far-off planet; and, finally, there was the problem of the long interval that would inevitably elapse when sending signals from the probes back to Earth. The probes were equipped with a large parabolic antenna, behind which two hexagonal housings were positioned: in one of these eleven scientific instruments, vital to the expedition's success, were installed and which were to investigate magnetic fields, cosmic rays, the Solar wind and its interaction with the interplanetary magnetic field. Once it was sufficiently close to Jupiter, the probe was to analyze the planet's environment and surroundings as well as its numerous moons; it was to take photographs; measure the gravitational field strength, and analyze the structure of the various atmospheres and ionospheres. Pioneer 10 was launched in March 1972 and soon accomplished its tasks; once it had photographed Jupiter for the first time at the minimum distance of 130,000 km (81,000 miles), it left the outermost limits of the Solar System in June 1983: it was the first object built by mankind to travel beyond this fabled boundary and it

Above: An artist's impression of a Pioneer probe during its voyage through interplanetary space, reaching beyond the orbit of Mars for the first time.

Below: This diagram shows the trajectory followed by Pioneer 11 after its launch from Earth on April 5, 1973. After a fly-by of Jupiter and Saturn the probe exited the Solar System, in common with its twin, Pioneer 10.

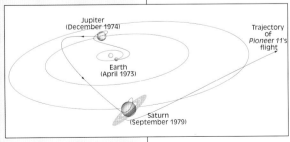

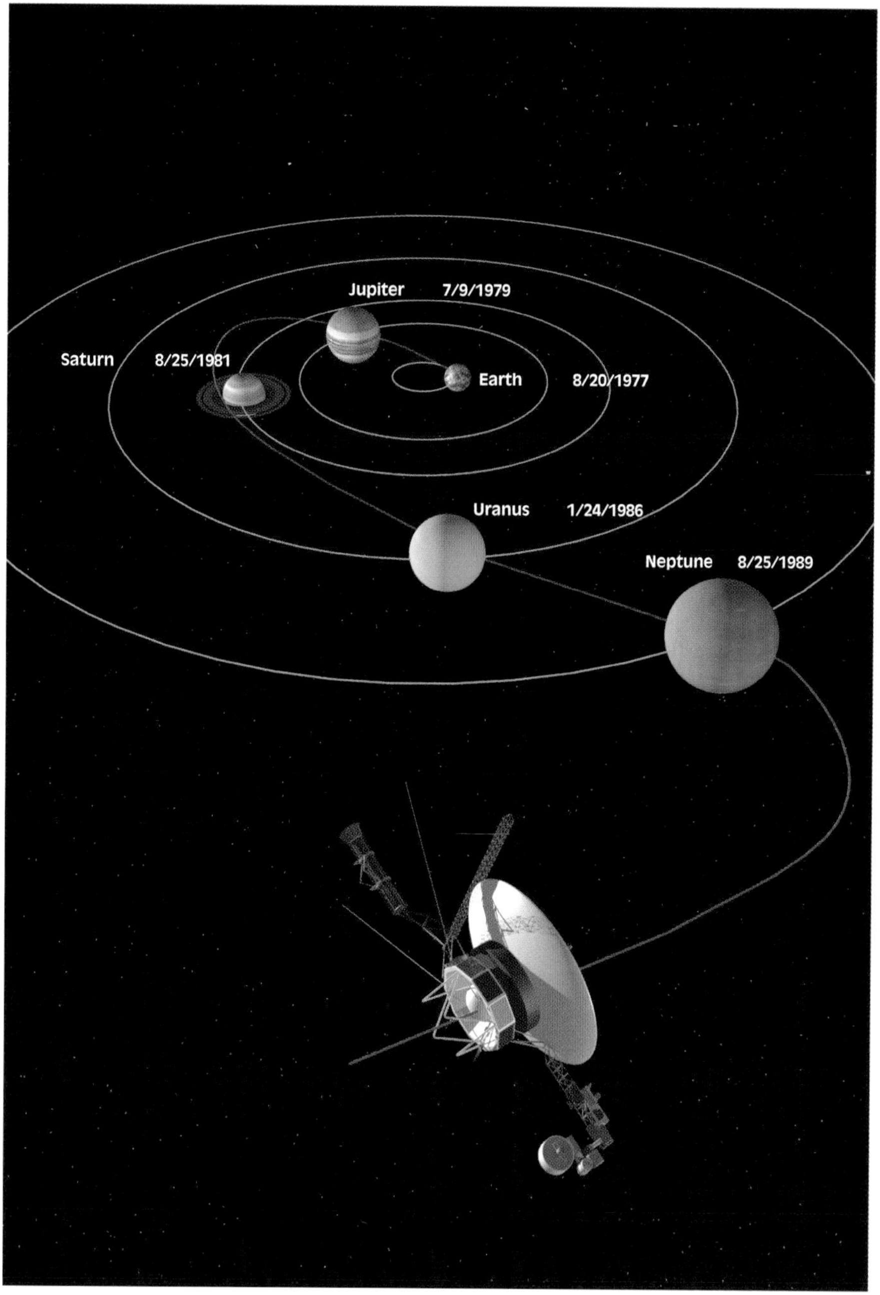

Jupiter 7/9/1979

Saturn 8/25/1981

Earth 8/20/1977

Uranus 1/24/1986

Neptune 8/25/1989

is estimated that it will reach Proxima Centauri in 26,000 years' time.

Pioneer 11 *was launched in April 1973 and, after fly-bys of Jupiter and Saturn, it left the Solar System in the opposite direction to that taken by its twin probe. Apart from their close-up reconnaissance of Jupiter, these probes also analyzed the giant planet's magnetosphere, measured the magnetic spectrum, the angular distribution, and the position of the magnetic electrons and of protons. In 1977 the U.S. Congress voted for a new planetary exploration project that had been submitted by NASA. This had been planned after the launch of the* Pioneer *probes and shortly before the* Voyager *missions were due to end. The project envisaged a mission to Jupiter with the aim of studying its complex magnetosphere and its atmosphere, using a module carrying instruments that could collect data direct. This mission was called* Galileo, *in honor of the discoverer of Jupiter's moons and the project included construction of a cosmic robot with a power-plant and fuel reserves to enable it to carry out various maneuvers. This was also the first launch of an interplanetary probe using a space shuttle. The* Galileo *probe was endowed with state-of-the-art systems, controlled by 19 micro-computers as well as by a single computer; this was to give it greater flexibility in carrying out its tasks and to enable data to be sent back to Earth at a rate of 134,000 bytes per second. The probe's instrument payload included radiometers, spectrometers and a camera that was 100,000 times more sensitive than the version used by the* Voyager *probes.* Galileo *was launched in October 1989, and exploited the gravitational fields of Venus and the Earth to complete two orbits around the Sun in order to acquire the gravitational thrust necessary to enable it to reach Jupiter in 1995.* Galileo *carried out several reconnaissance flights around Jupiter's moons. The probe also carried a module which it released in July of that same year; this module parachuted down into the atmosphere of Jupiter, analyzing its composition as it descended and transmitted data back to Earth for nearly one hour before it was destroyed. The* Galileo *mission had come to the end of its projected operational life by 1997, but NASA decided to extend it, re-naming it the* Galileo-Europa Mission, *with the aim of investigating liquid oceans on the moon Europa. The* Galileo Millennium Mission *plans to destroy the probe in September 2003 by having it plunge into Jupiter's atmosphere; the resulting friction on the probe will cause it to disintegrate.*

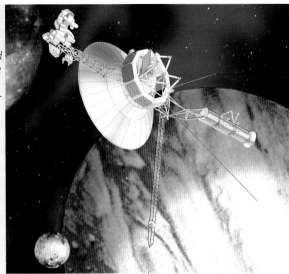

Above: A photo-montage showing the Voyager 2 probe between Jupiter and its moons as they were photographed by the probe's cameras in March, 1979.

Below: The launch of NASA's Voyager 2 probe from Cape Canaveral on August 20, 1977.

Facing page: Voyager 2's epic journey.

SATURN ♄

Above: An image of Saturn taken with the Hubble telescope's Wide Field and Planetary camera.

Below: The comparative dimensions of Saturn and Earth, showing their respective rotational axes in relation to the perpendicular of the ecliptic.

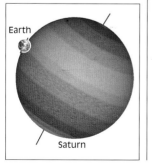

Earth

Saturn

Physical characteristics

Although we now know that each of the four giant planets possesses a ring system, Saturn is unique in the complexity and the wealth of detail of its rings. With a mass equivalent to approximately 95 times that of the Earth and a radius more than 9.45 times the terrestrial radius, Saturn is somewhat smaller than Jupiter but it is also the least dense planet of the entire Solar System, with a value of 0.69 times the density of water as compared to 1.25 times for Jupiter. Saturn rotates around its own axis once every 10.23 hours. Its mean distance from the Sun is 9.54 astronomical units, with an orbital inclination of barely 2.5 degrees in relation to the ecliptic, and it takes 29.46 years to achieve a complete revolution around the Sun. Because of its rapid rotation, Saturn is appreciably flattened at its poles, like Jupiter; in addition, its equatorial rotational period is 10 hours 13 minutes, while at higher latitudes it is 10 hours 38 minutes. It is difficult to say where the atmosphere begins on Saturn, since pressure and temperature vary with depth in such a way that there is no clear separation between regions in which the molecular hydrogen and helium are in a gassy state and those in which they change to a liquid state.

The complexity of the system formed by this planet, by its satellites and by its rings is such that it provides observers with a very exciting object of investigation in the fields of celestial mechanics and dynamics.

The rapid rotation of the planet and the presence of liquid metallic hydrogen have resulted in the generation of a dynamo effect, which in turn has produced a magnetic field and a magnetosphere which are second only to Jupiter's in their intensity and extent. Apart from the aesthetic beauty and number of its rings, Saturn is also of additional interest in that it has a satellite, Titan, with a dense atmosphere which some scientists believe could allow very complex organic molecules to exist, similar to those from which life on Earth originated.

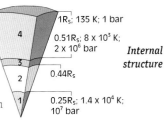

Internal structure

1—Core of silicates, minerals and ice
2—Metallic hydrogen
3—Transition zone
4—Molecular hydrogen

Regions

In common with Jupiter, Saturn is surrounded by a series of bands or belts parallel with its equator which differ slightly in color and are much more subtly shaded. These can be divided into the polar regions, temperate and tropical zones and equatorial belts. The northern polar region extends from 55° to 90° in the northern latitudes and varies in color, from fairly bright to very dark; the same applies to the southern polar region, extending between 70° to 90° in the southern latitudes.

The two temperate zones, north and south, stretch between 40° and 70° north and south respectively and their colors are generally bright but with few details. The northern temperate belt, at a latitude of 40°, is one of the most active zones. The southern temperate belt extends symmetrically in relation to the Equator. The northern tropical zone, contained within the latitudes of 20° and 40° north and the corresponding southern zone, between latitudes 20° and 40° south, are brilliantly colored and

Atmosphere

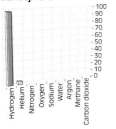

Characteristics of Saturn

Mean distance from Sun (AU)	9.54	Mass (g)	5.688 x 10^{29}
Mean distance from Sun (10^6 km)	1426.98	Mass (Earth = 1)	95.181
Orbital period (days)	10759.5	Equatorial radius (km)	60,268
Mean orbital velocity (km/s)	9.67	Equatorial radius (Earth = 1)	9.449
Orbital eccentricity	0.0556	Mean density (g/gm^3)	0.69
Apparent mean diameter of Sun	3' 22"	Mean density (Earth = 1)	0.13
Inclination of orbit to ecliptic (°)	2.488	Volume (Earth = 1)	761.449
Number of satellites	18	Ellipticity*	0.098

Equatorial surface gravity (m/s^2)	9.05
Equatorial surface gravity (Earth = 1)	0.93
Equatorial escape velocity (km/s)	35.5
Sidereal rotation period at equator	10 h 13 min 23 s
Inclination of equator to orbit (°)	26.73

*Ellipticity is (Re—Rp)/Re, where Re and Rp are the planet's equatorial and polar radii, respectively.

are bordered by dark-colored bands. Finally, there is the equatorial belt, stretching between 20° north and 20° south, in which many details are observable, as well as occasional whitish spots. The northern equatorial belt, between 0° and 20° north, is generally dark and fairly active; the southern equatorial belt, between 0° and 20° south, is also dark.

Atmosphere

Since Saturn has no solid surface, zero altitude is defined as the point at which the temperature reverses its progression. This inversion point also exists within Earth's atmosphere, at an altitude of approximately 30 km (20 miles) in the stratosphere, where the ozone layer is found. The inversion is caused by the ozone layer's absorption of solar ultraviolet radiation. The same phenomenon occurs in Saturn's atmosphere, but here methane is the absorbent gas. At altitudes below the zero fixed as a point of reference, the temperature increases with depth, and by approximately 100 K to approximately 300 K at a depth of 300 km (about 180 miles). Above the reference point zero, the temperature starts to rise again until it reaches 150 K at 200 km (120 miles) above it. Saturn's atmosphere consists of 96.3% molecular hydrogen, 3.3% helium and 0.4% methane. Saturn's temperature is not sufficiently cold to allow the methane to condense. Instead, at approximately 100 km (60 miles) below the reference point, where pressure is approximately 1 atmosphere,

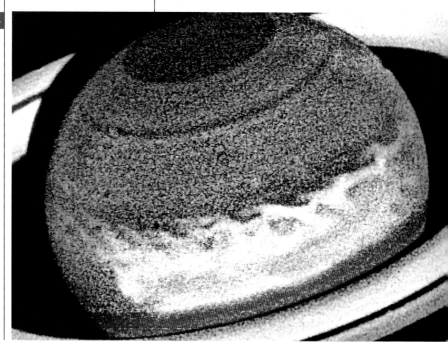

Below: An image of Saturn taken by the Hubble space telescope. An extensive whitish formation is visible along the equator.

Left: An image of Saturn taken by Voyager 1 on October 30, 1980.

Below: A photo-montage assembled by NASA's Jet Propulsion Laboratory with images obtained by the Voyager probes.

ammonia can condense and it is this which probably causes the whitish spots that have been observed.

Temperature

The increase in temperature, rising gradually towards a maximum above the zero reference point, is caused, as has already been explained, by the absorption of sunlight by the methane gas and also by the hydrocarbon fog. Conversely, below the reference point, where the lower the level, the more the temperature increases, the main agent is the heat of Saturn itself, released from its internal sources. In fact, Saturn, like Jupiter, radiates double the quantity of energy in infrared than it receives from the Sun. It is unclear what the source of this extra heat is, but it is thought that it originates from the separation of hydrogen and helium in its interior. Helium, as the heavier, would fall, freeing heat as it is compressed.

The heat freed in this way would be taken up to a level corresponding to the zero altitude of reference where it could escape freely into space. The transport of heat would occur through convection, that is to say, by means of the same mechanism by which water, when placed in a saucepan and put on a heat source, boils: the heavier currents of cold water descend towards the bottom of the pan, while the hot currents, being lighter, rise towards the surface. The same thing happens to the gas in the atmospheres of Saturn, Jupiter, Earth, Uranus, and Neptune.

Clouds and Winds

One of the main differences between the appearance of Jupiter and that of Saturn is the scarcity of distinct cloud systems in the atmosphere of the latter. This is probably due to the lower temperatures and lower gravity; as a result of these factors clouds form at a lower level and are partly

obscured by a fog of hydrocarbons which makes the colors of Saturn's belts paler and more blurred compared with those of Jupiter.

Atmospheric circulation on Saturn, as on Jupiter, is governed by various causes which differ from those affecting Earth. The difference in temperature at the poles and at the equator is very small, approximately

Above: A close-up of the rings of Saturn and the bright limb of the planet itself. This clearly shows how much more complex their structure proved to be than indicated by previous observations. They were also shown to be more numerous.

Facing page, below: An image of Saturn in false colors taken by the Very Large Array radio telescope in New Mexico, at a wavelength of 6 cm. The blanket of ammonia clouds is visible. The hotter regions are red, the cooler ones, in blue. The red band at a latitude of 35° N, with little ammonia content, makes it possible to glimpse a lower, warmer atmospheric layer underneath.

5°C (41°F), whereas on Earth the great difference between polar and equatorial temperatures is one of the main influences on terrestrial meteorology. In addition, the heat source that determines weather conditions and the seasons on Earth is sunlight, while on Saturn, as on Jupiter, an internal source of heat probably plays a more important role. The rapid rotation of the two giant planets has, moreover, a very significant influence on the dynamics of their atmospheres. Apart from these differences, many of the features that can be observed on Saturn's surface are analogous to terrestrial meteorological phenomena, such as cyclonic and anti-cyclonic formations.

The *Voyager* space probes have discovered a great deal about Saturn's meteorology. Most notably, it was possible to measure the speed at which the winds blow on this planet. The wind speeds are greatest at the equator and they blow in an easterly direction at a speed of nearly 500 m/s (1,600 feet/s); zonal winds also occur at both northern and southern latitudes of 30°, 50° and 60° respectively, with speeds ranging from 100 to 150 m/s (about 300-500 feet/s). Winds blow in an easterly direction at northern and southern latitudes of approximately 40°, 58° and 70° and

these, as is the case on Jupiter, are indicators of regions of instability, although general meteorological conditions seem to remain stable for very long periods.

What is surprising is the perfect symmetry of the wind distribution in the two hemispheres, despite the planet's equatorial inclination at 29° to the orbital plane; this should result in a seasonal effect in the two hemispheres. This state of affairs shows that at Saturn and Jupiter's distance from the Sun, solar radiation is too weak to influence the movement of the winds and the cause needs to be sought in the planet's internal heat.

Spots

Although the color contrast on Saturn is much less definite than on Jupiter, a number of oval spots occur which, like Jupiter's Great Red Spot, can remain visible for years and even for centuries. In the northern hemisphere at a latitude of 27° north, a spot has been observed which is mainly visible in the ultraviolet; since this radiation is absorbed to a greater extent than those with a longer wavelength, the conclusion is that the spot must be at a greater altitude than the clouds surrounding it. At a latitude of 42° north, three brown spots have been observed; the largest of which measures 5,000 x 3,300 km (3,100 x 2,000 miles). The largest stable oval spot, which has been given the name of "Big Bertha," is situated at a latitude of 72°; it measures 10,000 x 6,000 km

Above: These drawings by Galileo show Saturn as it appeared to him on July 30, 1610, September 3, 1616 and in October 1616. The first depicts a "trigeminal" Saturn, or triple planet, as it appeared to the famous Italian astronomer during his first observations, at a time when the rings had a "closed" appearance. The second and third drawings were made in 1612 after the temporary "disappearance" of the ring, when it had reappeared, in the form of fairly wide-open "handles." The third drawing clearly shows how close Galileo's records of his observations came to the real shape of the ring.

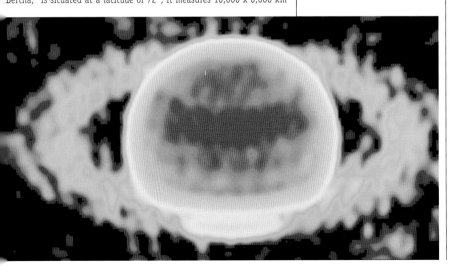

THE MOST BEAUTIFUL PLANETARY SYSTEM

During the Pioneer-Saturn missions of the 1970s and the Voyager missions of the 1980s, not only was a great deal of data on the planet collected but also some amazing images were obtained, all of which led to more in-depth studies of the planet's satellites and the discovery, in 1990, of a moon called Pan. Another great chapter in the story of the exploration of Saturn is also unfolding as a result of the Cassini/Huygens mission, a joint project achieved by NASA in collaboration with the European Space Agency (ESA) and the Italian Space Agency (ASI), this last agency having built the parabolic antenna for communication with terrestrial stations. The mission was launched in 1997 and reached its nearest approach to Jupiter in December 2000. Among other tasks, it carried out observations of the Jovian atmosphere, magnetic field and interactions with its satellites, until March 2001.

After a voyage at speeds of 58,000 km/h (36,000 miles/h) and three gravity-assist maneuvers (approaches to planets to exploit their gravitational attraction, which on this mission involves two gravity-assists of Venus and one of Earth), the probe will arrive in Saturn's vicinity in 2004, firing its rockets to insert itself into orbit around the planet. The orbital reconnaissance phase will last four years. The Huygens capsule will be released to parachute down through Titan's dense atmosphere and land on its surface: if this surface turns out to be liquid, the capsule will be able to float. It is thought that Titan could be in a pre-biotic state, similar to that which prevailed on Earth in the primordial era.

The capsule is expected be able to survive for anything from 3 to 45 minutes and its data will be picked up by the Cassini probe and sent back to Earth. The mother-probe will then complete 75 different orbits of Saturn over a period of four years, enabling it to study the planet and its moons from different angles.

Below: At 4.43 a.m. on October 15, 1997, the Titan IVB/Centaur rocket took off from Cape Canaveral, carrying the Cassini probe which was a joint NASA, ESA (European Space Agency) and ASI (Italian Space Agency) project. During its journey of approximately 2.2 billion kilometers (1.4 billion miles), which involves two orbits around Venus and one around the Earth in order to gather speed, the probe is expected to reach the orbit of Saturn in 2004, where it will spend four years gathering data.

Right: Saturn and its rings, photographed from the Catalina Observatory in the United States, on March 11, 1974. The bands parallel to the equator, similar to those of Jupiter, are discernible.

(6,000 x 3,700 miles). In the northern polar regions cyclonic phenomena have been observed which probably have their energy source in the latent heat freed by condensation of the water into ice under the visible layer of the ammonia clouds. The upward movement of the clouds is influenced by the planet's internal heat.

At sufficiently high altitudes the ammonia liquefies and falls again in the form of precipitation or rain. In this way a cycle of currents is established that prolongs the life of the clouds. Another great spot, reddish in color, is called "Anna's Spot," and is found at a latitude of 55° south; this spot measures 5,000 x 3,000 km (about 3,000 x 2,000 miles). Its color is due to the presence of phosphorus. Saturn's clouds do, in fact, contain a large quantity of phosphine (a molecule composed of one atom of phosphorus and three atoms of hydrogen) from which the phosphorus is freed when this substance is exposed to sunlight. Its resemblance to Jupiter's more extensive Great Red Spot leads scientists to believe that these formations represent a feature that is common to the atmospheres of the giant planets. Long-lasting vortices or whirlwinds also form in the terrestrial

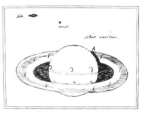

Above: Huygens, who observed Saturn with Cassini in 1675 from Paris, made this drawing showing the planet as it appeared through the high power telescopes made by Campani, which were unsurpassed in quality for instruments of their type. The double ring can be seen and also the dark belt on the planet's surface; to the right side of the sphere can be seen the shadow cast by the planet on the ring. Many years were to elapse, after the use of these superb telescopes was discontinued, before such clear observations of Saturn were repeated.

Left: A series of photographs of Saturn taken from the Lowell Observatory on various dates. The rings of Saturn lie on a plane that forms an angle of 27° with its orbital plane and as a result the appearance of these rings changes due to their varying aspect relative to the line of sight as the planet rotates; when the rings are seen edgewise (as in the top right photograph) they become almost invisible.

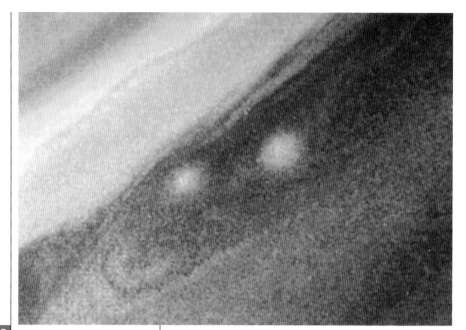

Above: A detail of Saturn's northern hemisphere in a photograph taken by the Voyager 2 probe on August 19, 1981.

atmosphere, lasting for some months and sometimes for years, but they are usually associated with mountainous formations, or they occur above the line of separation between *terra firma* and the ocean. There are neither mountains nor oceans on Saturn or Jupiter.

Vortices

A possible explanation could be that this is a case of a "solitary wave" which consists of a wave with a single crest, rather than the series of maxima and minima typical of light or sound waves or the waves of the sea. It has also been suggested that the vortex is linked with a much deeper and more vertically stable system, which enables it to survive even strong perturbations. One of the features noted from the time of the very first telescopic observations of Jupiter and Saturn is the presence of the parallel belts at the equator and oval spots of different colors. These, as we have already described, are more distinct on Jupiter, less distinct on Saturn.

The coloring of these spots depends on the properties of the substances present in the atmosphere and on the way in which sunlight excites and scatters or disassociates the atmospheric molecules. The layers of whitish clouds are the highest and their color is due to the presence of ammonia molecules; a little lower down a brownish color is visible, due to the presence of ammonium hydrosulphide, and still lower down there are pale blue clouds of water droplets. The Hubble Space Telescope has revealed that there is an extensive white formation that stretches out around the equator and which encircles virtually the entire hemisphere. From its color, it can be deduced that this is an enormous formation of ammonia clouds. It is thought that such a structure could be the result of a rare meteorological event, the cause of which is unknown, but it could also be a stable formation, usually hidden by the fogs of hydrocarbons and by the ammonia clouds. Another possible explanation is linked to the fact that a massive quantity of gas containing water vapor and ammonia, originating from hotter, lower zones could have been projected into the upper atmosphere. Once it had reached the highest part of the atmosphere, the gas would grow cold, forming fine, white ice crystals.

Internal structure

Saturn's internal structure is very similar to that of Jupiter. At the upper edge of its core the temperature is 12,000 K and pressure is 8 million atmospheres. For the purposes of comparison, these reach 30,000 K and 100 million atmospheres respectively on Jupiter. The size of Saturn's core is comparable to that of the Earth and its composition is probably rocky. The core is surrounded by liquid metallic hydrogen which extends outwards to a region that is estimated to occur at approximately 32,000 km (20,000 miles) under the cloud layer. In this cloud layer the temperature falls to 9,000 K and pressure falls to 3 million atmospheres. Still higher, an ocean of molecular hydrogen and helium is encountered and, finally, an atmosphere with complicated structures of cloud formations.

Below: The rings of Saturn observed from a distance of approximately 1.5 million kilometers (930,000 miles) by Voyager 1. The outermost ring (Ring A) and the innermost ring (Ring B) are separated by a wide, dark band known as the Cassini Division, discovered in 1675. Inside Ring A a narrow dark band is discernible, known as the Encke Division. Ring C, even further in, is a faint bluish color and this indicates different reflective properties; compared with the A and B rings, Ring C is more transparent. On the outside edge, Ring F is barely discernible.

THE DISCOVERY OF SATURN'S RINGS

Galileo was the first astronomer to observe Saturn, in 1610, when he sighted its disk with, on either side of it, two secondary bodies which became invisible during the years that followed only to reappear again, looking like "handles." In 1656 Huygens observed a shadow projected onto the surface of the planet and made out a ring shape, enabling him to provide an explanation, based on its inclination to the ecliptic, of the phases of its visibility: Saturn is surrounded by a thin ring, inclined at 28° on the orbital plane of the planet.

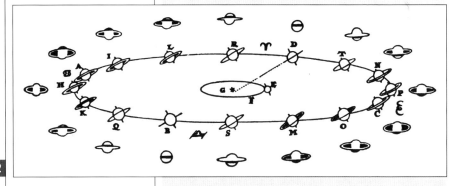

Above: The system of Saturn, according to Huygens. Depending on the planet's position in its orbit, aspects of the rings vary when observed from Earth.

The aspect of this ring when observed from Earth varies according to Saturn's orbital position; this means that at certain times the ring appears edge-on and vanishes from terrestrial observers' sight, only to reappear, gradually reaching its maximum inclination with the ring system wide open. In 1664 Campani spotted a dark belt as well as a lighter one in the outer zone of the ring; in 1675 Cassini observed the division that was named after him. The instruments of the nineteenth century made it possible to make out further divisions in the ring, such as the one named after Johann Franz Encke, in 1838. In 1850 another, very faint, semi-transparent ring, called the Crepe or Dusky Ring, was spotted, through which the body of the planet could be clearly seen inside the bright rings.

Astronomers were then able to deduce, from the shadows projected onto the planet's disk and from the fact that stars and planets disappeared when they passed behind the outermost ring, that this was very opaque and, bearing in mind its mass, it had to consist of fragments of material orbiting around the planet. This was confirmed by the discovery that the innermost zones of the ring rotated around the planet in 7 hours, while the outermost rings took 22 hours. The sub-division of the ring into ringlets was attributed to resonance phenomena originating in perturbations of the planet's satellites, which bring about rapid changes in the orbital periods of those particles thought to be present in the rings.

Magnetic Field

The presence of a liquid and conductive inner layer, like that of metallic hydrogen, led scientists to expect that Saturn would have its own magnetic field, produced by a dynamo effect. It was only in 1979 that its existence was confirmed by the *Pioneer 11* probe. The *Voyager* probes confirmed *Pioneer*'s discovery and measured the intensity of the field, which is approximately equal to the terrestrial field and 20 times weaker than that of Jupiter. This is probably due to the fact that the layer of metallic hydrogen is much less extensive than the Jovian layer.

Furthermore, the axis of the magnetic field is inclined by barely 1° in relation to the rotational axis, unlike Jupiter and the Earth, where the inclination is approximately 10°. Like Jupiter, Saturn is also a radio source, although only a very weak one. The planet emits radio bursts of wavelengths varying from a few hundred meters to several kilometers, with a periodicity of 10 hours 39.4 minutes, this time lapse is considered to be the planet's "real" rotational period. It is, in fact, constant, contradicting inferences drawn from the spots, which also possess their own longitudinal motion.

Magnetosphere

Like its magnetic field, Saturn's magnetosphere is mid-way in size between that of the Earth and that of Jupiter. It extends behind the planet, with the shape of a windsock. Where the solar wind meets the magnetic field at supersonic speed, what is known as a "bow shock" is

Below, right: Ring F observed by Voyager 1 from a distance of 750,000 km (470,000 miles). It is approximately 100 km (60 miles) wide and inside it numerous discontinuities, braids, knots and twisted rings are discernible, each approximately 10 km (6 miles) wide.

Below, left: A false-color image of the ring system. Computer processing exaggerates the color difference between the various rings in order to achieve detail enhancement.

formed and the solar particles change direction abruptly, sliding along the boundary surface that separates the magnetosphere from external space. The shock wave forms at a mean distance of approximately 1,800,000 km (1,100,000 miles) from the planet, while the magnetosphere is at a mean distance of 500,000 km (310,000 miles). Mean distance is the term used since these distances depend on the intensity of the solar wind. Between the magnetosphere and the shock wave is the

Resonance

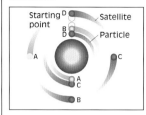

Curved waves

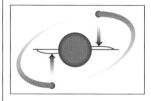

Shepherd satellite

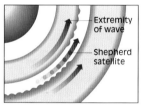

Above: A diagrammatic illustration of how Saturn's rings are formed: a magnificent example of celestial mechanics.

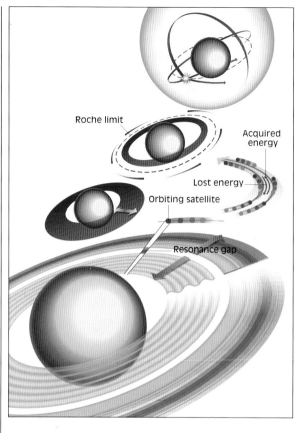

magnetopause. The nearest, and largest satellites, such as Titan, Rhea, Dione, Tethys, Enceladus and Mimas, as well as the rings themselves, have an effect on the magnetosphere that is far from negligible. For instance, Titan is the only one of Saturn's satellites that possesses an atmosphere, and emits molecules that can be neutral, or ionized which interact with the solar particles, creating a torus of hydrogen which extends inwards for approximately 700 km (450 miles) from Titan's orbit, until it reaches that of Rhea. Within a distance of approximately 400,000 km (250,000 miles) from Saturn is a torus of ionized oxygen and hydrogen atoms. These charged particles, moving in a spiral along the force lines of the field contribute to the growth of the local field and gather speeds, or energy values, which correspond to temperatures of half a million degrees Kelvin. The ionized hydrogen of the magnetosphere is absorbed by the inner satellites and by the rings. As is the case on Earth, the charged particles penetrate through the magnetic poles into the atmosphere, excite its gases and produce the phenomenon of aurora borealis as a result of this process. It was assumed that a similar phenomenon takes place on Jupiter and Saturn and the *Voyager* probes have, in fact, observed aurora borealis on both planets. The aurora borealis on Saturn are approximately 10 times weaker than those on Jupiter.

Rings

The rings of Saturn are more reflective than the planet's disk, which has an albedo of approximately 0.5, a little higher than that of the Earth which is 0.39 and similar to that of Jupiter. As a result the rings make a considerable contribution to the brightness of the planet which looks far more brilliant when the rings are viewed frontally as opposed to when they are seen edgewise. These rings, which lie on Saturn's equatorial plane, are seen edgewise by terrestrial observers at intervals of 13.75 and 15.75 years; this difference is due to the eccentricity of the planet's orbit. When observed edgewise the rings are barely visible, even with large, powerful telescopes, because they are very thin. The rings are denoted by capital letters. The three main rings are A (the outermost one), B and C. Rings A and B are brighter and separated by the "Cassini Division," named after the astronomer who discovered them. Ring C is semi-transparent and is consequently also known under the name of the Crepe or Dusky Ring. Ring A has a width of 14,600 km (9,000 miles) and is therefore larger than the Earth's diameter. Ring B is approximately 25,600 km (16,000 miles) wide, while the Cassini Division is much narrower than the rings, reaching a width of only 4,600 km (2,900 miles). Further in is Ring D, which is almost 7,000 km (4,400 miles) in width. The other rings, E, F and G, are all outside Ring A. Ring D does not have a clearly-defined inner edge, and this is therefore not distinct from Saturn's cloud tops; it has a brightness equal only to 1/100 of the dark ring C. Ring F is only 300 km (180 miles) wide. Between rings A and F, approximately 3,400 km (2,100 miles) apart, two satellites orbit the planet: Atlas and Prometheus. Even further out is ring G, which is approximately 3,000 km (1,900 miles) wide. Between rings F and G, separated by

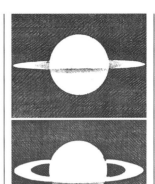

Above: Saturn's ring in two drawings by Huygens dating from 1656 and 1657 respectively. He has shown the shadow projected on the planet's disk, a detail that enabled Huygens to predict the true annular shape of the structure.

Below: Saturn's magnetosphere.
1. *Solar wind*
2. *Bow shock*
3. *Magnetopause*
4. *Force lines of magnetic field*
5. *Zone of intense radiation*
6. *Rotational axis*

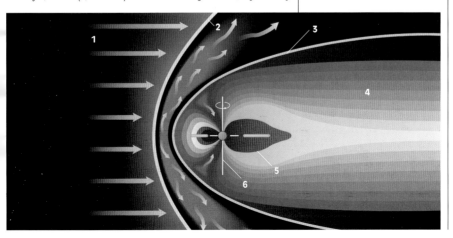

Above: The rings of Saturn: three photographs taken by Slipher at the Flagstaff Observatory in Arizona.

Below: Saturn and its rings photographed from the Catalina Observatory, United States, on March 11, 1974. It is possible to make out the bands parallel to the equator, similar to those of Jupiter.

almost 30,000 km (19,000 miles), three satellites orbit the planet: Pandora, Epimetheus and Janus. Outside Ring G, Mimas, a satellite that is slightly larger than the previous ones traces its orbit. Finally, there is the outermost ring, E, which is almost 302,000 km (188,000 miles) wide, a little less than the distance between the Earth and the Moon. The orbit of the satellite Enceladus lies inside ring E. Rings E, F and G are very faint, practically transparent. Unlike the larger rings, however, which have a vertical extension of approximately 200 meters (700 feet), E extends vertically for 2,000 km (1,200 miles) and G for 100 km (60 miles). Ring F is quite exceptional, with its braided structure and radial spokes.

The Shepherd Satellites

The two satellites of Prometheus and Pandora, the former slightly closer to Saturn than Ring F and the other outside it, are also known as the "shepherd" satellites because they focus particles towards the ring: the external satellite attracts the errant, outermost particles of the ring, slowing them down, while the inner satellite speeds up particles and returns them to the main ring zone. The ring is, moreover, formed from at least five components which move along orbits that intersect, giving rise to what is usually described as a "braiding" of the whole structure which changes the shape of the ring itself over times that vary from hours to months. It is probable that the two shepherd satellites trigger this variability in the shape of the ring. Of the three larger rings, ring C has the smallest mass and is composed of particles in excess of 1 centimeter in diameter, as was revealed by radio-astronomical observations. The color of the rings is reddish and this is perhaps due to the presence of dust containing ferrous oxide. Ring A has a mass mid-way between those of rings C and B. Ring B has the greatest mass, containing approximately three-quarters of the total mass of the ring system. Also, unlike the other rings, it is completely opaque. Ring B's albedo implies that it is almost entirely composed of ice particles.

Composition of the rings

The rings of Saturn are formed by thousands of thin rings or ringlets and by as many narrow divisions which give the whole structure the appearance of an old-fashioned gramophone disk. Before the *Voyager* probes' observations it was believed that these rings were formed by gravitational disturbances caused by Saturn's satellites. Something similar happens in the belt of asteroids orbiting between Mars and Jupiter; in this there are zones without any asteroids and these occur precisely where the orbital period would be an exact fraction of that of Jupiter. In the same way, the gravitational action of the satellites creates zones in which the particles cannot have stable orbits and they therefore undergo shifting. If these instabilities of gravitational origin can explain the larger divisions, they are not sufficient to account for the extremely complicated structure revealed by space probes. The most widely-accepted explanation nowadays involves the possible formation of spiral density waves caused by resonance between the particles of the rings and the outermost satellite of Saturn, Iapetus. It is not, however, yet known with any certainty whether the smaller rings and their divisions are permanent elements or not. Only future space missions that are planned to explore the environs of Saturn will be able to provide a definite answer. Nevertheless, ten years after the explorations carried out by *Voyager 1* and *2*, some details remain to be analyzed in depth from the images that they sent back to the terrestrial stations. It was only after exhaustive analysis of over 30,000 images sent by *Voyager 2* that an unexpected discovery was made: that of the existence of a moon, which is also the only one to fall within the system of the three main rings. Called Pan, this satellite is tiny, with a radius of approximately 10 km (6 miles).

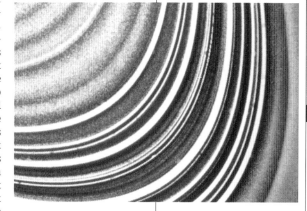

Above: Details of the innumerable divisions of Saturn's rings.

The origin of the rings

Before it was discovered that rings were not peculiar to Saturn and that they were present around all four giant planets, theories as to their origin were put forward specifically for Saturn. Although these theories can be adapted, at least in part, to relate to the other planets, the fact remains that only Saturn has such an extensive and highly-developed system. The French mathematician, Edouard Roche, formulated the theory that a satellite having the same density as its planet could not orbit it at a distance of less than 2.44 radii of the planet without being destroyed by tidal forces which the latter exerts. Nevertheless, the innermost ring, observed in the nineteenth century occurs at a distance of 2.3

times the radius of Saturn, and therefore within the Roche limit. In 1859, however, the Scottish physicist James Clerk Maxwell, demonstrated that the instability would apply only to a ring formed from liquid substances or from a solid aggregate and not to a ring composed of independent particles. It was therefore thought that these rings must have formed from debris from a large asteroid attracted by Saturn and shattered by its gravitational action, greater in the case of the nearest ring and less for the rings further away from the planet. Results from data collected by space probes have, however, shown that small satellites, of less than 100 km (60 miles) in diameter, can survive within the Roche Limit. This applies in the case of Saturn's satellites of Atlas, Prometheus, Pandora and Pan. Other hypotheses have been formulated: one theory is that the rings could be the residue of the material from which Saturn condensed, another that these rings formed at a later stage than Saturn, as a result of the disintegration of a fairly large satellite, of a size in the

Above: Further details of the countless divisions in the planet's rings.

order of the dimensions of Mimas, when it encroached within the Roche Limit. This last hypothesis is the most probable, because the total mass of all the rings is close to that of Mimas; in addition, the second of these hypotheses fails to explain why Saturn should have left such a quantity of residual material during its formation in this form, far greater than is the case with the other giant planets.

"Spokes" or radial formations

Another completely unexpected discovery was made by the *Voyager 1* probe, and was subsequently confirmed by *Voyager 2*, during the course of their lengthy reconnaissance of the planet. They discovered that radial formations existed in Ring B. These radial spokes appear as alternatively darkish and bright, depending on whether light illuminates the ring from the front or from behind. The radial spokes are approximately 10,000 km (6,000 miles) long and 2,000 km (1,200 miles) wide. Their presence is completely incomprehensible, since the particles of the ring obey Kepler's Laws and therefore those nearer the planet have orbital speeds faster than those further away.

This begs the question as to how it could be possible for these radial structures to form since, among other puzzling aspects, they should be very transitory, but have been seen to persist for several hours. It is possible that the magnetic field that was measured by the probes could be responsible for the formation of these radial spokes. Measurements carried out by the *Pioneer* and *Voyager* probes showed that gravitational forces and erosion of the micrometeorites should destroy the rings within the course of approximately one hundred million years, a far shorter stretch of time than that of Saturn's age which is approximately 4.5 billion years.

At present, the most likely explanation for the formation of Saturn's rings is provided by the hypothesis that they formed as a result of the destruction of a small moon when it collided with an asteroid or a comet. In this scenario, the rings would therefore be temporary formations that would be recreated every time that such a collision takes place. We now know that all four giant planets: Jupiter, Saturn Neptune and Uranus, have their own ring system, but it is difficult to understand why Saturn alone possesses such a complex system, with rings that are so distinct and visible. The theory of the destruction of a small satellite is a plausible one, but this begs the question as to why this did not happen in the case of Jupiter, a planet that also has many satellites in its system.

Above: Details of Ring A observed by Voyager 1 at a distance of 740,000 km (466,000 miles). The faint F ring is clearly visible on the right side of the image, while on the left side can be seen the wide Cassini Division.

Below: An illustration depicting NASA's Cassini probe at the point where it has been inserted into orbit around Saturn and is releasing the Huygens module. The module will parachute down onto the surface of the moon Titan to study its features.

Satellites

Saturn is known to have 18 satellites, and it is probable that there are many more. Titan is the largest satellite, not only among those belonging to Saturn's system, but of the entire Solar System. With a radius of 2,575 km (1,600 miles), it is larger than Mercury. Next in size comes Rhea, with a 765 km- (475-mile) radius, followed by Iapetus (720 km/447 miles); Dione (560 km/348 miles); Tethys (525 km/326 miles); Enceladus (250 km/155 miles) and Mimas (195 km/121 miles). These seven satellites were already known to exist by the end of the eighteenth century. Hyperion and Phoebe were discovered during the nineteenth

Above: An artist's impression of the European Space Agency's Huygens module as it parachutes down to land on Titan.

Facing page, top: A diagrammatic illustration of Saturn's satellites showing their comparative sizes.

century, but all the remaining known smaller satellites have been revealed by the *Pioneer* and *Voyager* space probes. The first six of these, in order of increasing distance from Saturn are Mimas, Enceladus, Tethys, Dione, Rhea and Titan, their orbital radii ranging from 185,520 to 1,221,850 km (115,277–759,257 miles) with near-circular orbits lying in, or nearly in, the equatorial plane of Saturn which is also the plane on which the rings lie. Hyperion, at 1,481,000 km (920,000 miles) from the center of Saturn, possesses a decidedly elliptical orbit that nevertheless lies virtually in the planet's equatorial plane, whereas Iapetus, at a distance of 3,561,000 km (2,213,000 miles) from the planet's center, has a circular orbit but it is inclined by approximately 15° in relation to Saturn's equator. All these satellites rotate around Saturn in the same orbital direction and the same direction as the planet. One of the most distant satellites, Phoebe, 12,952,000 km (8,048,000 miles) away from Saturn (a distance 3.6 times greater than that of Iapetus) has a very elliptical orbit inclined at 60°. Phoebe's orbital motion around Saturn is retrograde. This leads to the supposition that this is not a true satellite, formed of the same material as that from which the planet condensed, but is more likely to be an asteroid captured by Saturn's gravitational field. The other minor satellites are irregular in shape and their radii range from 10 to 100 km (6 to 60 miles), while their distances from the

Saturn's Satellites

Satellites	Mean distance from Saturn (km)	Distance in R_s	Orbital period (days)	Orbital inclination (°)
Pan	133,900	2.222	0.579	
Atlas	137,640	2.284	0.602	0
Prometheus	139,350	2.312	0.613	0
Pandora	141,700	2.351	0.629	0
Epimetheus	151,422	2.512	0,694	0.34
Janus	151,472	2.513	0.695	0.14
Mimas	185,520	3.078	0.942	1.53
Enceladus	238,020	3.949	1.370	0.02
Tethys	294,660	4.889	1.888	1.09
Telesto	294,660	4.89	1.888	0
Calypso	294,660	4.89	1.888	0
Dione	377,400	6.26	2.737	0.02
Helene	377,400	6.26	2.737	0.2
Rhea	527,040	8.74	4.518	0.35
Titan	1,221,850	20.27	15.945	0.33
Hyperion	1,481,000	24.57	21.277	0.43
Iapetus	3,561,300	59.09	79.331	14.72
Phoebe	12,952,000	214.91	550.480	175.3

center of Saturn range from 133,900 to 151,472 km (83,200 to 94,100 miles). Five of these small satellites have orbits that coincide with one another and with those of larger satellites. These are Epimetheus and Janus, both sited at a mean distance from the planet of 151,000 km (94,000 miles); Telesto and Calypso, both move in the same orbit as Tethys; and Helene, in the same orbit as Dione. These small satellites have a high albedo, ranging between 0.5 and 0.9. Their high reflective power suggests that they mainly consist of ice. The exception is Phoebe which is dark-colored, probably rocky, with an albedo of 0.06. In this respect Phoebe bears more of a resemblance to the minor satellites of Jupiter and Uranus, all of which have low reflectivity. Unlike the large satellites, all of which are more or less spherical, or ellipsoids, the small satellites, because of their irregular shape, are more like asteroids or comets' nuclei. This is because these small celestial bodies are subject to collisions and to being broken up, and their gravitational force is too weak to determine their shape, whereas the large satellites' shapes are determined by their own force of gravity and by their rotational motion. Most of the small satellites have rotational periods equal to their revolutionary periods around the planet, and as a result they always have the same face turned towards it, just as the Moon always has the same hemisphere facing the Earth.

- · Pan
- · Atlas
- · Prometheus
- · Pandora
- · Epimetheus
- · Janus
- · Mimas
- · Enceladus
- ● Tethys
- · Telesto
- · Calypso
- ● Dione
- · Helene
- ● Rhea

 Titan

- · Hyperion
- ● Iapetus
- · Phoebe

Orbital eccentricity	Radius (km)	Mass (g)	Mass (Moon = 1)	Mean density (g/cm³)	Mean density (Moon = 1)
	5				
0	20 x 15 x 15				
0	70 x 50 x 40				
0	55 x 42 x 35				
0.01	70 x 57 x 50				
0.01	110 x 80 x 80				
0.02	195	3.80×10^{22}	0.001	1.17	0.35
0.00	250	8.40×10^{22}	0.001	1.24	0.371
0.00	525	7.55×10^{23}	0.01	1.24	0.377
0	12 x 11 x 10				
0	15 x 12 x 10				
0.00	560	1.05×10^{24}	0.014	1.44	0.431
0.01	18 x 17 x 17				
0.00	764	2.49×10^{24}	0.034	1.33	0.398
0.03	2575	1.35×10^{26}	1.837	1.88	0.563
0.10	175 x 130 x 100				
0.03	720	1.88×10^{24}	0.026	1.21	0.362
0.16	115 x 110 x 105				

Above: This drawing, dating back to the 1930s, illustrates Saturn's system as it appeared when observed with a powerful telescope better than any photograph which inevitably entails the over-exposure of the planet if its satellites are to be clearly photographed.

Below: A detail of the surface of Mimas observed by Voyager 1 on November 12, 1980.

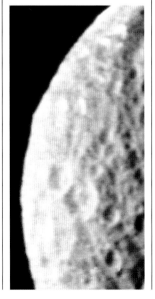

Major satellites

In the case of Saturn, the description of major satellites means those satellites whose radius is in excess of 100 km (60 miles) and which were already known before the space probe missions. For clarity, it is best to describe them individually, taking them in order of distance from the planet.

Mimas

Mimas, orbits at 185,520 km (115,280 miles) from Saturn. This satellite has a radius of 195 km (122 miles) and a density equal to 1.17 times that of water. It has the same face always turned towards Saturn, in common with all the other satellites, the one exception being Phoebe. Its surface is peppered by craters, dominated by the enormous Herschel crater, 130 km (81 miles) in diameter, huge when considered in the context of the satellite's size. The meteorite that created this impact crater must have been large enough to risk shattering Mimas completely. This satellite's density implies the presence of a small rocky core covered by a thick layer of ice.

Enceladus

Enceladus orbits at a distance from Saturn of 238,020 km (147,910 miles); it has a radius of 250 km (155 miles) and a density equal to 1.24 times that of water. Like Mimas, it must consist predominantly of ice, with a small rocky core. Its albedo is near 1, reflecting nearly all the incident light. Difference types of terrain have been observed on its surface; ridges, smooth and uneven plains, and areas covered with craters. The most striking feature is an extensive plain, virtually free of craters but which is traversed by long grooves or "ditches." This distinctive surface morphology of Enceladus may be due to the fact that the satellite Dione, which is much larger and more dense and has a period twice that of Enceladus, may exert tidal effects sufficient to keep the interior of Enceladus relatively hot. Material may be expelled from the interior in the form of soft ice, leveling the surface.

Tethys

Tethys orbits 295,000 km (183,300 miles) from Saturn and two small satellites, Telesto and Calypso follow the same orbital path, Telesto 60° ahead and Calypso 60° behind. These three satellites are therefore sited on the vertices of an equilateral triangle, in accordance with a well known law of celestial mechanics. The radius of Tethys is 525 km (325 miles) and its density is equal to 1.24 times that of water, similar, therefore, to that of Enceladus. The main feature of its surface is the presence of a large crater, 400 km (250 miles) in diameter, only a little less that the satellite's radius. Another unique feature is the presence of an enormous trench, Ithaca Chasma, which crosses virtually the entire sphere, from the north pole almost as far as the satellite's south pole; it is 100 km (60 miles) wide and 4–5 km (2½–3 miles) deep. Perhaps, when Tethys had not long been formed, its interior was liquid and covered with an icy crust. The progressive freezing of the interior could have caused it to expand and produced this enormous fracture. This, however, does not explain why only one, huge fracture formed rather than many smaller ones.

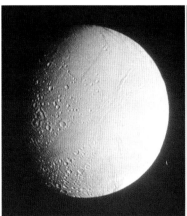

Dione

Dione orbits at a distance of 377,000 km (234,000 miles) from Saturn and has a radius of 560 km (350 miles). It is somewhat denser than the preceding satellites, at 1.44 times the density of water. Dione probably has a rocky core that is larger than the other satellites' nuclei; it has a lower albedo than Saturn's other moons. But it is interesting to note that the albedo of Dione's surface is anything but uniform: it varies from a value of 0.6 on the leading hemisphere (i.e. the hemisphere that leads as the planet rotates), and only 0.3 on the trailing hemisphere. Dione's most curious feature is a structure called Amata, which has a diameter of 240 km (150 miles) and could be an irregular crater or a basin. Most of Dione's craters have diameters of less than 30 km (18 miles). Near the south pole there is a long valley which extends, almost in a straight line, for approximately 500 km (300 miles).

Rhea

Rhea orbits at a distance of 527,000 km (327,000 miles) from Saturn and its radius measures 764 km (475 miles): it is the second satellite after Titan. Its density is equivalent to 1.33 times that of water. In common with Dione, its leading hemisphere differs in appearance from the trailing hemisphere. The former is even and flat, the latter darker and scattered with whispy features. Its surface, like those of the other satellites,

Above, left: An image of the satellite Enceladus, taken by Voyager 2 *on August 25, 1981.*

Above, right: An image of Tethys taken by Voyager 2 *on August 25, 1981.*

Below: Dione against the background of Saturn.

is covered with ice and north of the equator it is particularly liberally pitted with craters.

Titan

Titan is not only the largest of Saturn's satellites, but it is by far the largest satellite in the Solar System, with dimensions even greater than the planet Mercury. Its radius is 2,575 km (1,600 miles). Titan orbits at a distance of 1,222,000 km (759,000 miles) from Saturn. It has a density

Right: A detail of the surface of Rhea, imaged by Voyager 1 *on November 12, 1980.*

Below: An image of Titan, taken on August 25, 1981 by Voyager 2.

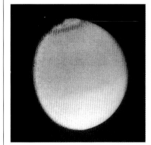

equal to 1.88 times that of water and must therefore consist of at least half rocky material and half water-ice. The escape velocity of Titan is 2.5 km/s (1.6 miles per second), which leads to the assumption that it must be able to retain volatile elements, since the low temperatures encountered at this distance from the Sun slow down the movements of atoms and molecules to speeds lower than the escape velocity. These predictions were confirmed in 1944 by spectroscopic observations which indicated the presence of methane in a gaseous state. The *Voyager 1* probe measured the composition of Titan's atmosphere and revealed that it contained 90% nitrogen, with methane as a minor constituent and, perhaps, argon. It also detected the existence of much smaller quantities of ethane, acetylene, propane and other organic compounds. The composition of Titan's atmosphere resembles that of Earth in certain respects. In fact these are the only bodies in the Solar System whose atmospheres are rich in nitrogen. Surface-level pressure is 1.5 atmospheres and the temperature is 95 K (or 178°C/288°F below zero).

The temperature falls to 70 K at several tens of kilometers above the surface, only to rise again to approximately 200 K in the satellite's stratosphere. Titan's color oscillates between reddish and bright orange. The southern hemisphere is uniformly bright while the northern hemisphere is darker and redder. The meteorology of Titan is very complex, and showers of methane and nitrogen probably occur. It is also possible that oceans and rivers of liquid methane exist on its surface, and hills of solid methane.

Iapetus

Iapetus, at a distance of 3,561,000 km (2,213,000 miles) from Saturn and with a radius of 720 km (450 miles), is the outermost of the large satellites. Its most notable feature is the extremely low albedo of the leading hemisphere during rotation, with a value of 0.04 while the trailing hemi-

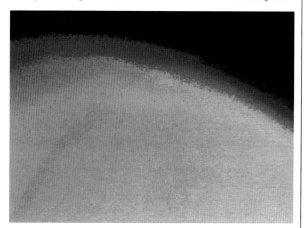

sphere is very bright with an albedo of 0.5. Because this satellite's density is equal to 1.21 times that of water, Iapetus is probably covered with ice. It is thought, however, that the dark hemisphere may be covered with dust expelled by the surface of Phoebe following micrometeorite impacts. This material, shifting with a spiral movement towards Saturn could have come into contact with the dark hemisphere of Iapetus. Two objections can, however, be made to this hypothesis: the material from Phoebe is different in nature from that of the dark side of Iapetus; furthermore, at the terminator or border separating the bright and dark areas there is a ring of dark material, 400 km (248 miles) in diameter, which cannot come from a source external to Iapetus.

Hyperion and Phoebe

In size, Hyperion and Phoebe are mid-way between the larger and the smaller satellites. Both are very irregular in shape. The three semi-axes of Hyperion which orbits at 1,481,000 km (920,200 miles) from Saturn are, respectively, 175, 130 and 100 km (108, 81 and 60 miles); those of Phoebe are 115, 110 and 105 km (71, 68 and 65 miles).

Above, left: The limb of Titan as observed by Voyager 1 *on November 1, 1980. The thick veil of bright blue mist and smog surrounding the satellite is very obvious.*

Above, right: An image of Iapetus taken by Voyager 2 *on August 22, 1981.*

URANUS

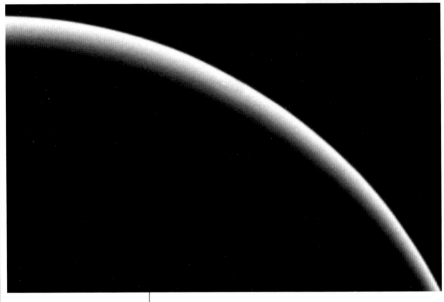

Above: In the upper layers of the atmosphere of Uranus, hydrocarbon fogs occur above large methane cloud belts, while in the lower layers, clouds of ammonia and water are thought to be present. It is due to the methane's absorption of red and bright orange light that the planet acquires its characteristic greenish-blue color.

Physical characteristics

Uranus is, so to speak, beyond the pillars of Hercules in the Solar System. Before the invention of the telescope, Saturn was the most far-off planet known to exist. From the time of its discovery, which happened on March 13, 1781, until 1986 when *Voyager 2* approached it, very little was known about its physical structure and chemical composition, and even less about its magnetic field. There was even uncertainty about its rotational period, estimates for this varying between 11 hours and approximately 24 hours. Uranus is the only planet whose rings have been discovered through use of sophisticated observational techniques based on Earth.

Its most extraordinary feature, which makes it unique in the outer Solar System, is the tilt of its rotational axis, which is almost perpendicular to the plane of the ecliptic, which means that it alternately has its north pole and its south pole turned towards the Sun. The rotational axis of Uranus is tilted to such an extent, and at such a right angle that its motion is technically retrograde or anti-clockwise and what should be its north pole actually lies under the ecliptic.

The convention is, therefore, to express the inclination of Uranus as 98° instead of 82°. The tilt of this axis has been determined by that of the orbital planes of the major satellites, discovered by telescopic observation, and these have not revealed any orbital perturbation caused by the planet's equatorial bulge. There would be appreciable perturbations if the orbits of the satellites lay outside the equatorial plane.

Origin of the inclination of the rotational axis

The fact that the orbits of all the satellites of Uranus lie in the equatorial plane of the planet, despite its extraordinary inclination to the ecliptic, indicates that it is very probable that they were formed from a proto-planetary disk, which was closely connected to the planet at such time when its obliquity assumed its present value, perhaps as a result of some cataclysmic event. A less convincing hypothesis but one that cannot be excluded is that a series of reciprocal gravitational perturbations between the planet and its satellites caused them to orbit in the equatorial plane itself.

The orbital period of Uranus is 84 years and as a result one of its two poles can be observed from Earth every 42 years. This strange anomaly marks Uranus out from the other planets, presents us with an unresolved element in all the theories that have been formulated in the attempt to explain the origin and formation of the Solar System: in fact there is no model that can satisfactorily account for its rotational axial inclination, without adducing a collision with another body. Apart from this puzzling feature, Uranus resembles Neptune in size and density. It differs from the latter, however, in that it does not have its own internal heat source: this absence is another enigma. As a result of the *Voyager 2* probe's observations, the rotational period of Uranus has been confirmed as 17 hours

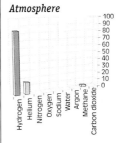

Internal structure

1R_U; 80 K; 1 bar
0.7R_U; 2.5 x 10^3 K; 2 x 10^5 bar
0.3R_U; 7 x 10^3 K; 6 x 10^6 bar

1—Rocky core
2—Envelope of molecular hydrogen, helium and ammonia
3—Mantle of water-ice, methane and ammonia

Atmosphere

Hydrogen, Helium, Nitrogen, Oxygen, Sodium, Water, Argon, Methane, Carbon dioxide

Characteristics of Uranus

Mean distance from Sun (AU)	19.19	Mass (g)	8.684 x 10^{28}
Mean distance from Sun (10^6 km)	2870.99	Mass (Earth = 1)	14.531
Orbital period (days)	30685	Equatorial radius (km)	25,559
Mean orbital velocity (km/s)	6.81	Equatorial radius (Earth = 1)	4.007
Orbital eccentricity	0.0461	Mean density (g/cm^3)	1.29
Apparent mean diameter of Sun	1' 40"	Mean density (Earth = 1)	0.23
Inclination of orbit to ecliptic (°)	0.774	Volume (Earth = 1)	62.121
Number of satellites	15	Ellipticity*	0.00229

Equatorial surface gravity (m/s2)	7.77
Equatorial surface gravity (Earth = 1)	0.79
Equatorial escape velocity (km/s)	21.3
Sidereal rotation period at equator	17 h 12 min
Inclination of equator to orbit (°)	97.86

Ellipticity is (Re—Rp)/Re, where Re and Rp are the planet's equatorial and polar radii, respectively.

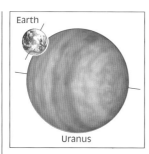

Above: A comparison of the size of Uranus and the Earth and of the tilt of their respective rotational axes.

Below: A false-color image of Uranus, taken by Voyager 2 on January 14, 1986 from a distance of 12.9 million km (8 million miles). The ring-shaped dark marks are not real features, but are caused by the presence of dust in the lens of the camera.

and 12 minutes. Its magnetic field is almost as strong as that of Saturn, with its axis almost perpendicular to the rotational axis. Uranus is blue in color and its disk is streaked by sporadic white clouds. The planet's coloring is caused by the presence of hydrogen and methane in its atmosphere: these absorb sunlight mainly in red and infrared, reflecting blue and green.

Atmosphere

In common with Jupiter and Saturn, Uranus also possesses an atmosphere that mainly consists of hydrogen and helium, 85% of which is accounted for by molecular hydrogen and 15% by helium. This atmosphere is very extensive and it is estimated that it accounts for 30% of the radius of the planet. It overlays an ocean of water, ammonia and methane at very high pressure and temperature, 200 atmospheres and 2,500 K respectively. Since it is very difficult to ascertain where the surface of Uranus begins, it is assumed, as a point of reference, that it is equal to the level at which minimum temperatures are recorded.

Voyager 2's observations made it possible to find out about variations in winds and temperature at different latitudes. Given Uranus's vast distance from the Sun and the consequent small quantity of radiation it receives, there is very little variation in temperature; as a result of the

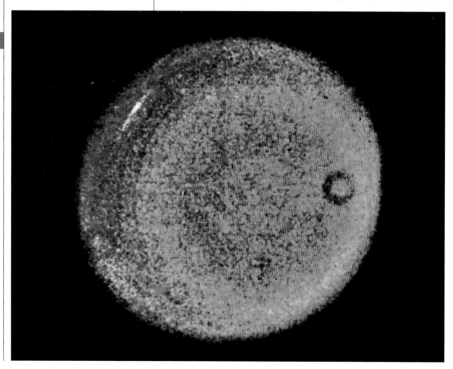

inclination of the rotational axis to the ecliptic, however, the poles of Uranus receive a higher quantity of sunlight than the planet's equator and the polar temperatures are higher.

The speed and latitude of the winds can be calculated from the movement of the clouds, as is the practice with Jupiter and Saturn. On Uranus, however, only a very few cloud formations are observed. It can be said that the winds are azimuthal: at mid-latitudes their speed approaches 180 m/s (590 feet/s) in the direction of the planet's rota-

 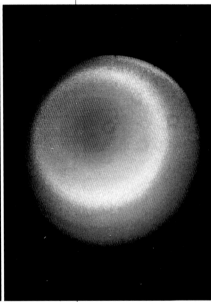

tion, while at low latitudes they blow in the opposite direction to that of the planet's rotation, at speeds up to 100 m/s (330 feet/s). Alone among the planets in the Solar System, the polar areas of Uranus receive a greater amount of sunlight than equatorial regions; it was believed that this peculiarity was a determining factor in establishing the circulation pattern of the winds and it was expected that on Neptune, which although very similar to Uranus in every other respect, has an equatorial inclination of 29° relative to the ecliptic, the wind regime would differ greatly. But these two planets behave in a virtually identical manner as to wind behavior and this has come as a complete surprise to the scientific community.

Above, left: Uranus as seen by Voyager 2; the bluish color is due to the absorption of the red component of sunlight by the methane gas present in the atmosphere.

Above, right: Voyager 2 shows the pole of Uranus. The computer-enhanced colors make it possible to discern a hazy area which stretches as far as the mid-latitudes.

Internal Structure

The density of Uranus is virtually the same as that of Jupiter. However, because of its mass, which is approximately 22 times lower, this planet, unlike Jupiter and Saturn, does not possess a layer of liquid metallic

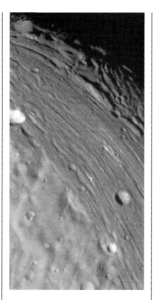

Top: A close-up of the surface of Miranda, one of the planet's moon observed by the Voyager 2 probe.

Above: Another image of Uranus.

Right: A diagram of the magnetic field of Uranus. The magnetic axis and the rotational axis are almost perpendicular.
1. Magnetic axis
2. Rotational axis
3. Magnetopause
4. Bow shock
5. Solar wind

hydrogen. The core of the planet has a radius of approximately 7,500 km (4,700 miles). This is surrounded by a mantle approximately 10,500 km (6,500 miles) deep, composed of molecular hydrogen, helium methane and ammonia in a liquid state (hence called its "ocean") and, this in turn is surrounded by a surface layer of hydrogen and helium which gradually merges into the atmosphere, and is approximately 7,600 km (4,700 miles) deep. A hypothesis has been put forward that the core is of a rocky consistency and mainly composed of silicate and iron; however, after some experiments were carried out under laboratory conditions and following processing of *Voyager 2*'s data, both the size and even the existence of a solid core is now in doubt.

The astrophysicist, W.B. Hubbard, prepared a substance composed of water, ammonia and isopropyl alcohol in his laboratory and subjected it to a pressure of 2 million atmospheres. He called this experiment "synthetic Uranus" because the properties of this liquid behave under pressure much like the environments found in the interiors of Uranus and Neptune.

Magnetic Field

It took a planetary mission to enable scientists to discover the magnetic field of Uranus. Its existence had, however, already been suspected, because the *IUE (International Ultraviolet Explorer)* astronomical satellite had measured strong, variable ultraviolet radiation which implied the presence of aurora borealis. The intensity of the magnetic field at surface level is slightly lower than that of the terrestrial field.

Another surprising fact about Uranus is that the axis of its magnetic field is inclined at 55° in relation to its rotational axis, whereas in the case of all the other planets that possess a magnetic field, the

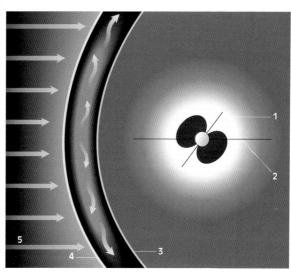

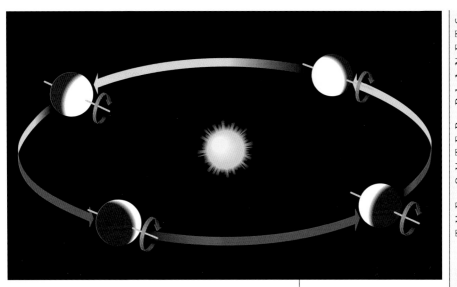

angle between the two axes never exceeds 10°. The magnetic poles therefore occur at intermediate latitudes, nearer to the equator than to the poles.

Since the magnetic axis is almost perpendicular to the plane of the ecliptic, as is the case with all the other planets that have a magnetic field, the magnetosphere of Uranus has no particularly unusual features. Nevertheless, because of the planet's rotation, the tail of the magnetosphere is twisted in the plane of the ecliptic. All the satellites and the rings are engulfed in the magnetosphere, which has the same rotational period as the planet. The high energy particles, that is to say the electrons and protons present in the magnetosphere, cause blackening of the organic matter present on the icy surfaces of the Uranian moons, which therefore appear very dark to observers.

The two exceptional features of Uranus: the inclination of its rotational axis, lying almost on the orbital plane, and the very wide angle between the magnetic axis and the rotational axis, are almost certainly linked. As is the case for Jupiter and Saturn, Uranus' magnetic field is the result of a dynamo effect, caused by the rapid rotation of the internal, fluid layers, with its axis more or less coinciding with the rotational axis.

Rings

The rings of Uranus, working from the innermost rings outwards, are U2R, with a radius varying from 37,000 to 39,500 km (23,000 to 24,500 miles), and consequently having an inner edge that occurs approximately 11,440 km (7,100 miles) from the surface of the planet; U6R, with a radius of 41,850 km (26,000 miles); U5R, with a radius of 42,420

Above: The rotational axis of Uranus lies almost on its orbital plane, so that Uranus can best be described as "rolling" along its orbit, rather than rotating. The alternating seasons are indicated in this diagram.

Below: An image of Uranus taken on January 14, 1986 showing the rotation of two cloud formations. The area observed is the north pole and the rotation is taking place in an anti-clockwise direction during a period of approximately 17 hours.

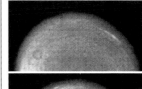

km (26,360 miles); and U4R, with a radius of 42,580 km (26,460 miles). These are followed by the rings designated a, b, h, g, d, l and e, the radii of which vary from 44,730 to 51,160 km (27,800 to 31,790 miles). The total width of the ring system amounts to 14,160 km (8,800 miles), equivalent to approximately half the radius of the planet. The larger rings had already been discovered in 1977 through observation of stellar occultations by Uranus, from the Kuiper Airborne Observatory, an airplane with a very high operational ceiling that was specially equipped to carry out astronomical observation. When a star is behind the disk of a planet, it is no long visible from Earth and it reappears when the planet, moving along its course, travels away from the star. During the occultations caused by Uranus, it was observed that both before and after the actual occultation, the occulted star's own light was partially hidden five times and this "winking" was repeated symmetrically, but in reverse order, after the occultation. It was evident that the star was transiting behind five partially absorbent objects, arranged symmetrically in relation to the diameter of the planet. The most likely explanation was that this must involve a ring system. From the duration and the depth of the minimum values of light it was possible to calculate the width and density of each ring. In general, the rings of Uranus are very thin and have very clearly-defined edges. All the rings, with the exception of h, are slightly eccentric. Ring e is 20 km (12 miles) wide at its minimum distance from the planet and 98 km (61 miles) at its maximum distance. Between the rings more extensive zones of diffused material overlap one another.

Features of the rings

The rings are formed of objects of varying sizes, varying from boulders in the order of 1 or 2 meters (3–6 feet) in size to particles approximating to a micrometer which account for less than 0.1%. The rings have a very low albedo, never exceeding the value of 0.15; this explains their low visibility. Ring e, between the two "shepherd" satellites 1986 U7 and 1986 U8, is the outermost ring and is also the widest. It has a very low albedo, reflecting only 0.05 of the incident sunlight, and has a complex structure in which three zones can be distinguished: the outermost, and brightest, which is 40 km (25 miles wide), followed by a darker zone of almost the same width, and finally an internal zone, only 15 km (9 miles) wide, but brighter.

The other rings, further in towards the planet, are all very dark and almost circular in shape. The tenth ring, discovered by *Voyager 2*, is very thin and occurs between rings e and d and is barely visible.

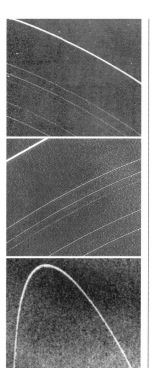

Above: The nine-ring system observed by Voyager 2 *from distances of: 2.52 million km (1.57 million miles) away, 1.12 million km (696,000 miles), and finally, from only 125,000 km (77,700 miles) away.*

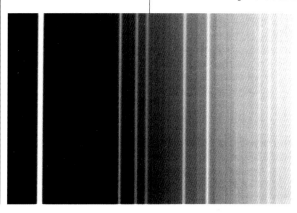

In addition to the ten rings, whose existence has been confirmed, there are almost certainly many others, less than approximately 100 kilometers (60 miles wide), but all difficult to spot, either because they are so thin, or because of their very dark color.

Satellites

Before the *Voyager 2* mission, Uranus was known to have 9 rings and 5 satellites. Apart from its discovery of the tenth, very thin ring, the probe increased the number of satellites to 15 of the now known 21. Those already known are, of course, the largest ones and also those furthest from the planet. They include Oberon, 582,600 km (362,000 miles) from Uranus and with a radius of 760 km (470 miles), followed by Titania, Umbriel, Ariel and, finally Miranda, at a distance of 129,780 km (80,650 miles), with a radius of 235 km (146 miles).

The density of these satellites varies from that of Ariel, which is equivalent to 1.66 times the density of water, to that of Miranda, equal to 1.35 times. All the other satellites are very small, with the exception of Puck which has a radius of 75 km (46 miles) and is also the outermost of the small satellites, at 86,010 km (53,450 miles) from Uranus. The smaller satellites including Belinda and Cordelia occur at distances ranging from 75,260 km (46,780 miles) and 49,750 km (30,900 miles) respectively.

All the satellites are outside the rings, with the exception of Cordelia which, in common with Ophelia which is at a distance of 53,760 km (33,400 miles), act as the pair of "shepherd" satellites to ring e. The larger satellites are similar in size to Saturn but they are denser and much darker. The brightest is Miranda, which reflects 30% of incident light; Oberon and Titania reflect 20%, while Umbriel only reflects 12%. The smaller satellites are much darker, with an albedo of approximately 7%. Their surfaces are covered with rocky material and ice.

The rotational period of the five large satellites is equal to their period of revolution, and as a result they always have the same face turned towards the planet. Nothing is known about the rotational period of the small satellites, but it is probable that they behave in the same way. *Voyager 2* sent images of the larger satellites back to Earth, with detail resolution of approximately 10 kilometers (6 miles).

Above: Oberon as observed from a distance of 660,000 km (410,000 miles). The smallest distinguishable details measure 12 km (7 miles).

Facing page, below: A false-color image of a region of the rings of Uranus.

- Cordelia
- Ophelia
- Bianca
- Cressida
- Desdemona
- Juliet
- Portia
- Rosalind
- Belinda
- Puck
- Miranda
- Ariel
- Umbriel
- Titania
- Oberon

Oberon

Voyager 2 carried out reconnaissance of Oberon's surface which is flat and dark, with some craters surrounded by lighter zones. A very large crater has been observed with a pale central peak and, in common with the other craters, a much darker floor. On the limb of the satellite a mountain approximately 6 km (4 miles) high has been observed, its altitude about one-tenth of the satellite's radius. This is truly impressive, bearing in mind as a comparison that Earth's highest mountain, Mount Everest, only rises to 0.14% of the terrestrial radius.

Titania

Nearer to Uranus, the second large satellite, Titania, has large paler areas on a fairly dark background and many impact craters. Its main feature is a visible fracture or trench near the terminator between the bright hemisphere and the dark hemisphere which indicates past tectonic activity. There are numerous valleys 50-100 km (30-60 miles) wide and several hundred kilometers long. When compared with the diameter of the satellite, which is 1,600 km (1,000 miles), these formations appear to be on a gigantic scale; such an uneven surface is probably the result of meteorite bombardments in the past and of subsequent tectonic activity.

Umbriel

Moving on past Titania further towards the planet, next comes Umbriel, which has a darker surface, less uneven than any of the other satellites. The most interesting feature is its large pale, ring-shaped feature, which

Satellites of Uranus

Satellites	Mean distance from Uranus (km)	Distance in R_U	Orbital period (days)	Orbital inclination (°)
Cordelia	49,750	1.95	0.335	0.14
Ophelia	53,760	2.1	0.276	0.09
Bianca	59,160	2.31	0.435	0.16
Cressida	61,770	2.42	0.464	0.04
Desdemona	62,660	2.45	0.474	0.16
Juliet	64,360	2.52	0.493	0.06
Portia	66,100	2.59	0.513	0.09
Rosalind	69,930	2.74	0.588	0.28
Belinda	75,260	2.94	0.624	0.03
Puck	86,010	3.37	0.762	0.31
Miranda	129,780	5.08	1.414	3.4
Ariel	191,240	7.48	2.52	0
Umbriel	265,970	10.41	4.144	0
Titania	435,840	17.05	8.706	0
Oberon	582,600	22.79	13.463	0

occurs near the equator and the diameter of which is approximately one-tenth of that of Umbriel itself.

Ariel

Still closer to Uranus is Ariel, its surface peppered with craters, and eroded into very wide valleys, the results of several types of activity. The craters were caused by impacts, while the fault scarps and graben indicate stresses acting on the crust, suggesting considerable internal activity. The branching valleys cut across one another and their deep floors contain winding gullies, which may have been eroded by liquid.

Miranda

Finally, we come to Miranda, which is the smallest of Uranus's major satellites and the nearest to the planet. One of the first images, photographed from a distance of 1.38 million kilometers (857,000 miles), shows an entire hemisphere, on which can be seen, almost in the center, a large, dark spot and, below it, a bright formation in the shape of the letter "V." In a photograph taken at closer range, showing almost the entire hemisphere, two types of terrain occur: one is paler and full of craters and the other is darker and furrowed by numerous grooves. In the images taken from a distance of 31,000 km (19,300 miles) a great deal of detail is visible, with more ancient terrain, scattered with craters, while another is completely criss-crossed by grooves and yet another, which is even more uneven. The V-shaped structure, when seen in greater close-up, is an area with many craters situated between two heavily grooved

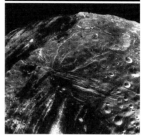

Above, top: Details of Ariel's surface taken from a distance of 130,000 km (81,000 miles). The smallest details visible measure 2.4 km (1.5 miles).

Above: Details of Miranda's chaotic surface.

Orbital eccentricity	Radius (km)	Massa (g)	Mass (Moon = 1)	Mean density (g/cm^3)	Mean density (Moon = 1)
0	15				
0	15				
0	20				
0	35				
0	30				
0	40				
0	55				
0	30				
0	35				
0	75				
0	235	6.89 x 10^{22}	0.001	1.35	0.4
0	580	1.26 x 10^{24}	0.017	1.66	0.5
0	585	1.33 x 10^{24}	0.018	1.51	0.45
0	790	3.48 x 10^{24}	0.047	1.68	0.5
0	760	3.03 x 10^{24}	0.041	1.58	0.47

zones. The high resolution of the images taken so close to the satellite means that it is possible to observe far smaller details and numerous geological structures which are already known to exist in other parts of the Solar System. Probably the individual structures belong to different geological eras. It is thought that the most ancient of these are those which have the greatest number of craters. As has already been noted in other bodies of the Solar System, the period of heavy bombardment dates back at least three and a half billion years. The impact craters are generally shallow, probably because of viscous relaxation of surface materials. Various stratigraphic systems can be identified, sometimes with clear separation between them, which suggest the presence of deep, thin fractures, out of which material flowed that led to the formation of the lower layers. The furrowing of the terrain, sometimes as deeply as several kilometers, reveals materials with differing albedos.

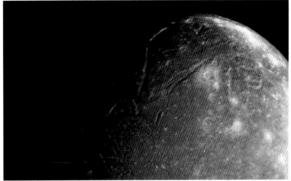

Top: This image shows numerous details of Miranda's surface. This was the satellite to which Voyager 2 came closest.

Above: Titania observed from a distance of 500,000 km (310,000 miles). Detail resolution is 9 km (5.6 miles).

Above, right: Ariel, as seen from a distance of 130,000 km (81,000 miles). Detail resolution is 2.4 km (1.5 miles).

Geology of the satellites

It is interesting to consider the ways in which the geological histories of the five large satellites appear to differ. The most distant and largest satellites, Oberon and Titania, do not show evidence of any great internal activity; Umbriel and Ariel, smaller in diameter and less distant from the planet, have surfaces that show signs of considerable internal activity; finally, Miranda, the smallest and nearest to the planet shows signs of a great deal of internal activity. The progressive increase in internal activity of the satellites, the closer they are to Uranus is probably evidence of an increase in tidal attraction exerted by the planet on its nearest satellites.

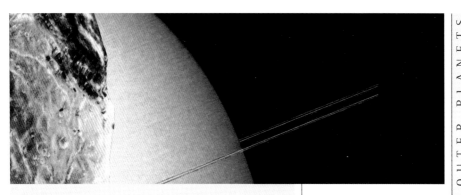

VOYAGER'S EPIC JOURNEY

Uranus was observed from a near distance in space terms for the first time in January 1986 when the Voyager 2 probe made a fly-by of the planet at a distance of only 81,500 km (50,600 miles).

Voyager 2 traveled at a speed of 15 km/s (9.3 miles per second) and transmitted the data it recorded back to terrestrial stations at a reduced rate of 21.6 bytes per second, in order to ensure that the data was accurately and safely received. The period allocated to observation of the planet was scheduled to end on February 25 and this allowed enough time for the terrestrial stations to receive thousands of photographs, showing what Uranus really looked like. It turned out to be enveloped in an atmosphere of helium and hydrogen, together with small quantities of acetylene and the methane which gives the planet its blue color in the photographs we can now see. The radio transmissions confirmed the existence of a magnetosphere, with a magnetotail extending for at least 10 million km (6 million miles); radiation was found to be of a similar intensity to that of Saturn. Ten moons were discovered, adding to the number of those already known and observations were made of the curious composition of a known satellite, named Miranda. Part of the surface of Miranda was shown to have been formed very recently, while the other part was extremely ancient: this leads scientists to suppose that this satellite is an aggregate of materials dating from different eras, following a violent impact. Two new rings were also discovered and ring arcs, some of which are only 50 meters (160 feet) wide. With this observational phase completed, the probe had to make use of its backup receiver: its instrument platform was partially jammed on the azimuthal plane but the probe nevertheless managed to continue on its journey as planned towards its final destination, Neptune, which it reached according to plan three and a half years later.

Above: A photo-montage showing the close-up of Miranda's surface against the background of the planet Uranus.

Below: A photo-montage that reconstructs the encounter of Voyager 2 with the planet in January 1986. The probe came to within a distance of 81,500 km (50,600 miles) to the planet and observation continued until February 25, 1986.

NEPTUNE

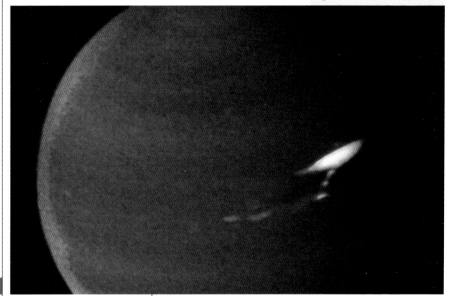

Above: An image of Neptune taken by Voyager 2. The colors are enchanced to show the great oval spot and the reddish atmosphere more clearly.

Physical characteristics

Neptune is slightly smaller than Uranus and there are many similarities between the two planets; Neptune's radius is 24,800 km (15,400 miles), its mass is equivalent to 17 times that of the Earth and its mean density is equal to 1.64 times that of water: all this data is very close to that of its neighbor. The same can be said of its rotational period which is 16.1 as compared with 17.2 for Uranus.

When observed from Earth, Neptune appears as a small disk with an angular diameter of 2", approximately half that of Uranus. While Uranus is situated at the limit of visibility by the naked eye, when Neptune appears, lit by the Sun, it is five times less brilliant: to put it another way, it would take five Neptunes to equal the brightness of Uranus. This explains why very little was known about Neptune prior to *Voyager 2*'s fly-by of the planet in 1989. The probe's observations confirmed its similarity to Uranus, including the fact that it is surrounded by a system of four rings and eight satellites.

Neptune is a sea-blue color, mainly due to the methane present in its atmosphere. The planet's disk is streaked with clouds, white, bright and frequently changing, which rotate more slowly than the planet. Some have a period of 16 hours and therefore, compared with the planet's rotation, their circulation is retrograde. On the other three large planets, the clouds rotate more rapidly than the planet they surround. It is not known why Neptune behaves differently in this respect. The method of stellar occultation which enabled the rings of Uranus to be discovered, was also used for Neptune. The results were, however, somewhat inconclusive. Out of approximately 20 occultations, 15 had unsatisfactory results, while the remaining occultations revealed a dipping of the stellar light before or after the occultation by the planet; not, however, repeated symmetrically at the beginning and at the end of the occultation, whereas this is the case with Uranus. Scientists believe that this may be connected with occultations by some small and unknown satellite; or that Neptune may not have a well-formed ring system like the other three giant planets, that instead it may have incomplete rings, best described as arcs.

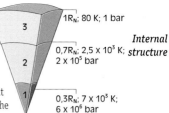

Internal structure

1—Rocky core
2—Envelope of molecular hydrogen, helium and ammonia
3—Mantle of water-ice and ammonia

Atmosphere

Neptune's atmosphere is very similar to that of Uranus. Even the temperature and pressure values at various levels, above and below the reference point, fixed at that level at which the temperature ceases to fall, are the same. The only difference is found at approximately 25 km (15 miles) above the reference level. At this altitude the temperature on Uranus remains constant, whereas that of Neptune starts to rise again and reaches 150 K at an altitude of 250 km (150 miles) above the reference level, where the pressure is approximately one ten-thousandth of an atmosphere. Even the chemical composition is approximately the same: 80% water vapor, helium, methane, water and carbon dioxide, which condense to form clouds and rain. Neptune's atmosphere contains

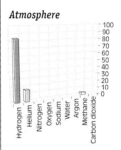

Atmosphere

Characteristics of Neptune

Mean distance from Sun (AU)	30.06	Mass (g)	1.024 x 10²⁹
Mean distance from Sun (10⁶ km)	4497.07	Mass (Earth = 1)	17.135
Orbital period (days)	60190	Equatorial radius (km)	24,764
Mean orbital velocity (km/s)	5.45	Equatorial radius (Earth = 1)	3.883
Orbital eccentricity	0.0097	Mean density (g/cm³)	1.64
Apparent mean diameter of Sun	1' 04"	Mean density (Earth = 1)	0.30
Inclination of orbit to ecliptic (°)	1.774	Volume (Earth = 1)	57.675
Number of satellites	8	Ellipticity*	0.017

Equatorial surface gravity (m/s²)	11.0
Equatorial surface gravity (Earth = 1)	1.12
Equatorial escape velocity (km/s)	23.3
Sidereal rotation period at equator	16 h 6 min
Inclination of equator to orbit (°)	29.6

*Ellipticity is (Re—Rp)/Re, where Re and Rp are the planet's equatorial and polar radii, respectively.

Above: An image of the planet Neptune.

Right: The Great Dark Spot is a vast region of high pressure, unmasked by the tropospheric methane clouds that hide the lower regions of the atmosphere.

Below: Images of the Great Dark Spot taken during successive rotations of Neptune. They shown changes to the Spot.

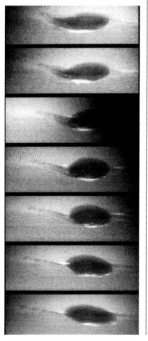

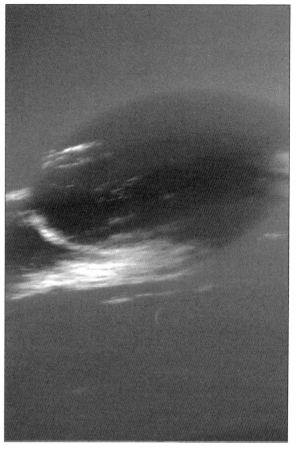

two cloud layers at different altitudes which differ in composition. At 80 km (50 miles) below the reference level, where the temperature is 130 K and pressure is 3 atmospheres, a layer of ammonia and hydrogen sulfide occurs; 40 km (25 miles) higher, where pressure is approximately 1 atmosphere, there is a layer of methane clouds.

At even higher altitudes, where pressure falls from a tenth to approximately one-thousandth of Earth's atmosphere, there are thin hazes of hydrocarbons, which result from the disassociation of methane caused by sunlight. These hydrocarbons combine together to form larger particles which descend into the warmer regions of the atmosphere, where they combine with hydrogen, to form methane again. This methane is again transported upwards by cloud columns, described as columnar clouds, and the cycle thus begins all over again.

THE MATHEMATICAL PLANET

Astronomers engaged in the study of Uranus during the nineteenth century realized that their calculations of its positions were inherently erroneous because they had not taken into account the effects of attraction by another, hitherto unknown celestial body which was sufficiently far away from the Sun as not to perturb the motion of Jupiter and Saturn, but near enough to Uranus to influence its movement. John Couch Adams in Cambridge and Urbain Le Verrier in Paris both embarked upon research into this unknown planet, using two different methods.

Adams fixed the position of the stars within the relevant zone in order to find out, when he checked again later, whether any of these had moved. Le Verrier collaborated with astronomers in Berlin who used a very accurate celestial atlas, the Akademische Sternkarte, informing them of the position where he expected the unknown planet to occur and asked them to check movements of stars in that area.

German astronomers started their research into this matter on September 23, 1846, and that very same evening Johann Galle noted the presence of a relatively bright star in a position where the map had nothing recorded. This was observed by Encke immediately afterwards using Berlin observatory's great equatorial telescope, and the new celestial body appeared in the form of a disk at only 55' away from the position indicated by Le Verrier and 1° 49' from the position calculated by Adams; it was also possible to observe its movement in relation to the stars in the background. This was one of the most significant discoveries made by mathematical astronomy during the nineteenth century.

A few weeks after the discovery of Neptune in 1846 William Lassell spotted the first satellite of the planet, later to be called Triton; the second, Nereid, which was much smaller and further away, was discovered by Gerard Kuiper in 1949.

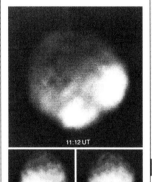

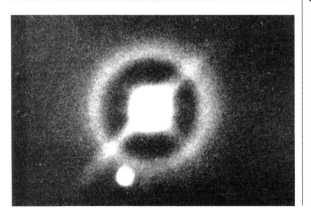

Left: Neptune and its satellite Triton in a photograph taken in 1949. The lengthy exposure needed to capture the image of the satellite on the photographic plate resulted in the planet being very over-exposed. The crossed ring was caused by elements of the telescope.

Above: Several images of Neptune, taken in succession from Earth. Before the planetary missions, it was impossible to achieve better-quality images.

Cloud layers

Neptune's clouds have a layered structure, clearly visible in the most successful photographs which show very high clouds casting their shadows on other, lower layers. A more reliable way of showing this stratification can be achieved by photographing several images at various wavelengths, ranging from ultraviolet to visible ultraviolet to infrared. Only the highest clouds are visible in ultraviolet light, which is scattered by the particles in the upper atmosphere. Observing methane's typical wavelengths, other details of the clouds can be made out including an edge that is brighter than the rest of the disk, caused by a layer of methane clouds situated at a high altitude.

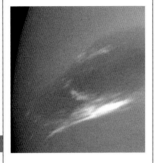
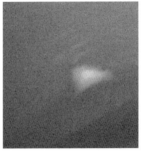

Above: Neptune's clouds at various wavelengths.

Below: A comparison of the size of Uranus and the Earth and of the angle of their respective rotational axes.

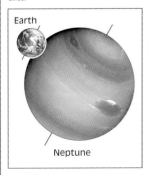

Spots and clouds

In common with Jupiter and Saturn, Neptune also exhibits oval spots. The most extensive of these is its Great Dark Spot, a cyclonic storm feature equal in size to that of Earth and approximately a quarter of Neptune's radius. This spot rotates in an anti-clockwise direction over a period of approximately 10 days. This enormous turbine of gas can be seen because it is not covered by the layer of methane clouds which cover almost the entire surface of the planet. Approximately 50 km (30 miles) higher up, whispy clouds that are similar to cirrus clouds continually form and dissolve. When Neptune's methane-rich atmosphere is forced to rise because of the pressure of the Great Dark Spot, it cools and the methane condenses, forming the clouds. To the south, there is another spot, the Dark Spot 2, above which there are ascending currents of methane which produce paler clouds. Apart from the cirrus-type clouds, those of the Great Dark Spot and the columnar clouds of the Dark Spot 2, another type of cloud has been observed, called scooter clouds because they move around the surface faster than the other cloud formations. Since these clouds rotate with the same period as the internal regions of Neptune, it is thought that they must originate in a deep, hot spot but all we can manage to observe is a rising column of hydrogen sulfide emanating from the cloud layer under the methane clouds.

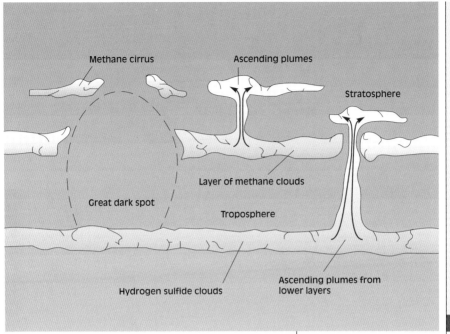

Above: Clouds in Neptune's atmosphere form at three levels. The highest are cirrus, which surround the planet; around the Great Spot ascending clouds have been observed; while lower down is a blanket of hydrogen sulfide.

Winds and currents

Neptune's atmosphere is much more active than that of Uranus, although they are very alike. During the reconnaissance carried out by the *Voyager 2* probe in the summer of 1989, the probe measured winds of speeds near to 600 m/s (2,000 feet/s), equivalent to 2,160 km/h (1,400 miles/h), blowing in a westward direction, in the opposite direction that is, to the planet's rotational direction. These are exceptionally violent winds, blowing at speeds almost ten times greater than the strongest terrestrial winds.

This begs the questions as to how winds of this intensity can blow on a planet that receives barely $1/900$ of the solar energy that reaches Earth. Neptune has its own heat source, as indicated by the fact that the planet emits a quantity of energy 2.7 times greater than it receives.

Perhaps it is precisely this internal heat source that accounts for the wind activity which, it must be remembered, is not present on Uranus. Uniquely among the giant planets, Uranus lacks its own source of internal heat. The transfer of heat from the interior towards the atmosphere entails the formation of warm, ascending currents and cold descending currents, caused by the transfer of heat from the interior of the planet outwards into the atmosphere.

Temperature

Despite the low temperatures on Neptune, when the planet is observed in infrared it appears very bright. Maps of the surface drawn up using the infrared instrument on board the *Voyager 2* probe, show that the temperature is a little higher at the equator and at the poles, but lower in the zones at intermediate latitudes. To put it more precisely, the infrared radiation image indicates that the zones of equal temperature are distributed in distinct belts parallel to the equator, and occur symmetrically in relation to the equator. There is an equatorial belt at a temperature of approximately 58 K with spots at 60 K; then there are two belts at latitudes approximately 15° above and below the equator, with temperatures of 57 K; two belts sited symmetrically in relation to the equator at latitudes of 30° with temperatures of 53–54 K; in another two, circumpolar, belts temperatures rise again to 57 K and, finally, come the poles with temperatures almost equal to those of the equatorial region. This state of affairs is puzzling, not least because during *Voyager 2*'s observations the Sun was at its zenith above the mid-latitudes. This reflects that Neptune's internal heat is more significant than the heat that it receives from the weak solar radiation. The interaction of these two heat sources may produce complex ascending and descending gas currents, which could explain this type of surface temperature distribution. To put it more succinctly: the gas currents at mid-latitudes rise upwards; they then cool and having reached areas over the poles and the equator, they fall once more

Above: A map of Neptune at a wavelength of 29 microns taken by Voyager 2. The colors indicate the temperatures (in Kelvin degrees) of the various zones. The Great Spot's temperature does not differ from that of the areas around it. The polar and equatorial regions are a little warmer than the other regions.

Right: A diagram showing how Neptune's magnetic axis is inclined at 47° in relation to the planet's rotational axis.

to lower altitudes and warm up again. Whereas on Jupiter the temperature of the Great Red Spot is lower than that of the rest of the belt in which it is situated, on Neptune no difference in temperature was detected by the infrared radiometer over the Great Dark Spot compared with regions surrounding it.

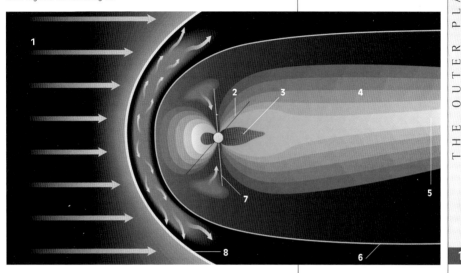

Magnetic Field

Eight days before *Voyager 2* reached its closest point to Neptune, the probe recorded short wave radio-waves emitted by the planet, which were repeated every 16.11 hours. This was proof that Neptune possesses a magnetosphere populated by charged particles which, moving within the magnetic field generated in the interior of the planet, produced radio-wave emissions. From this it could be reasoned that the periodicity of 16.11 hours represented the planet's rotational period, and this was subsequently confirmed by the details visible on its surface. In addition to these stronger, periodic bursts, it was discovered that there were a dozen weaker emissions at lower frequencies. Neptune's magnetic field, like that of Uranus, has produced some surprises. The first of these was that its magnetic axis is inclined at 47° to its rotational axis. In this respect Neptune again has much in common with Uranus, where the same angle measures approximately 55°. As a result the entire magnetic field is twisted into a spiral as Neptune rotates around its own axis. Secondly, the source of the magnetic field is not situated in the center of the planet, but is very displaced, almost half-way between the center and the surface. A somewhat similar situation is encountered on Uranus, although here it is much less exaggerated, because in Uranus's case the distance from the center is only one-third of its radius. These results suggest that the magnetic field of Neptune must be generated inside a fluid

Above: Neptune's magnetosphere
1. *Solar wind*
2. *Rotational axis*
3. *Zone of intense radiation*
4. *Force lines of magnetic field*
5. *Magnetotail*
6. *Magnetopause*
7. *Magnetic axis*
8. *Bow shock*

mantle which circulates around the solid core, more iron-rich than that of Uranus, which would explain the greater mean density of the planet. The strength of Neptune's magnetic field is equal to approximately 0.1 gauss, half that of Uranus.

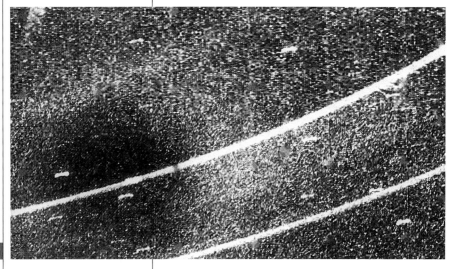

Above: A detail of Neptune's two main rings.

Rings

The rings of Neptune are very faint and have very distinctive breaks in them which have made observation of Neptune's stellar occultations puzzling. As has already been noted, the star's light occasionally faded before it disappeared behind the planet's disk, without however repeating this with a symmetrical fading on its emergence from behind the disk. This behavior suggested that Neptune was not surrounded by complete rings but by arcs, at least three in number. How these arcs could persist and not disperse was, however, a mystery.

What it suggested was the existence of at least six shepherd satellites, or the possibility of the existence of another large satellite nearby, such as Nereid which would have to describe a very tilted and/or very eccentric orbit. *Voyager 2* finally revealed the number and structure of the rings. The two large rings are 1989 N1R and 1989 N2R.

The first, outer ring is 62,900 km (39,100 miles) from the center of Neptune and is under 50 km (31 miles) wide. This is a single structure in which can be observed three denser regions: these are the arcs indicated by the stellar occultations. The other ring which consists of at least 40% dust, occurs 53,200 km (33,100 miles) from the center of Neptune and it is less than 15 km (9.3 miles) wide. In addition to these two main rings, *Voyager 2* discovered two more.

The first of these, which is much wider than the others, is situated between the two larger rings, the outer edge of which is at a distance of 59,000 km (37,000 miles) from the center of Neptune and, at its inner edge, 53,200 km (33,100 miles). It is an extremely faint ring, with a dust content of virtually nil. Its inner edge is formed by the second large ring, 1989 N2R. Within its radius, at 57,500 km (35,700 miles) from the center of the planet, there is another, brighter, ring which has been named 1989 N5R. Finally, there is a fourth ring, 1989 N3R, 41,900 km (26,000 miles) from the center of Neptune; this is the innermost ring and it is approximately 1,700 km (1,100 miles) wide. It, too, is also very faint, and consists of dust and gas.

Rings, arcs and dust

Voyager 2's revelations also indicate the existence of a disk of dust that would fill Neptune's entire internal system. It is estimated that the total dust content of the ring system must be approximately 100 times greater than in the Jovian system. The difficulty in observing Neptune's rings is also due to their low albedo. Methane, which at this distance from the Sun occurs in its solid state, is transformed into a series of dark-colored carbon compounds, because of the continual bombardment by solar radiation. The rings lie in the equatorial plane of Neptune and are immersed within the planet's magnetosphere. Because of the strong tilt of the magnetic axis in relation to the rotational axis, the rings traverse a wide strip across the magnetosphere in common with the satellites, while the planet rotates. It can therefore happen that at a given moment, the rings and moons are in the plane of the magnetic equator and after eight hours,

Above: The densest sections of Neptune's most extensive ring look bright, while the less dense sections are barely visible.

Below: The crescent of Triton and crescent of Neptune. Neptune looks white because its hazy atmosphere scatters all the light rays.

that is to say half a Neptunal day, they are positioned so that they transit over Neptune's magnetic poles. In this continual motion across the magnetosphere, the rings and satellites collect charged particles, and thus alter the structure of the planet's magnetosphere.

The persistence of dust in this system is a very short-lived phenomenon compared with the actual age of the planet. This means that a constant bombardment of meteorites must take place on Neptune if the rings are to be kept supplied with dust to replace that which they lose as a continuous process. The presence of arc-shaped structures, unchanged for at least five years, which is the length of time that has elapsed between the first terrestrially-based observations of the stellar occultations and the arrival of the *Voyager 2* probe in the vicinity of Neptune, forms one of the many still-unresolved problems of the Solar System.

Below: A diagrammatic illustration of Neptune's satellites, showing their comparative sizes and respective orbits.

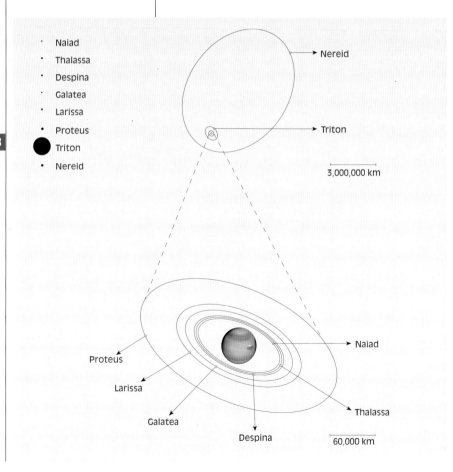

Satellites

Before the *Voyager 2* probe's mission, only two satellites of Neptune were known to exist: Triton, 354,800 km (220,500 miles) away from the center of Neptune, with a radius of 1,350 km (840 miles), and Nereid, at 5,513,400 km (3,426,000 miles), with a radius of 170 km (110 miles). Of the six new moons discovered by *Voyager 2*, Proteus has a radius of 200 km (120 miles), greater than that of Nereid, but it had never been observed from Earth because, being so close to Neptune, it was immersed in the planet's light. The other five satellites are all very small, their radii ranging from 95 km (60 miles) in the case of Larissa to 25 km (15 miles) for Naiad. The distances of Larissa and Naiad from the center of Neptune range from 73,600 km (45,700 miles) to 48,000 km (30,000 miles), respectively. The four innermost satellites, Naiad, Thalassa, Despina and Galatea, occur inside Neptune's ring system.

The six satellites discovered by *Voyager 2* follow virtually circular orbits with direct or prograde motion, almost coincident with the equatorial plane of Neptune, but Triton's motion is retrograde in relation to the planet's rotation, with a period equal to its orbital period. This means that Triton always has the same face turned towards Neptune. The orbit of Nereid is inclined at 29° in relation to the equatorial plane of the equator of the planet but is also very elongated, meaning that its distance from the planet varies from a maximum of 567,680 km (352,800 miles) to a minimum of 141,920 km (88,200 miles). The orbital features of these two outermost satellites of Neptune imply that they are bodies that have been captured by the planet.

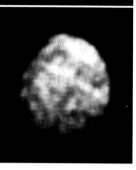

Satellites of Neptune

Satellites	Mean distance from Neptune (km)	Distance in R_N	Orbital period (days)	Orbital inclination (°)	Orbital eccentricity	Radius (km)
Naiad	48,000	1.94	0.3	0	0	25
Thalassa	50,000	2.02	0.312	4.5	0	40
Despina	52,500	2.12	0.333	0	0	90
Galatea	62,000	2.5	0.429	0	0	75
Larissa	73,600	2.97	0.554	1.212	0	95
Proteus	117,600	4.75	1.212	0	0	200
Triton	354,800	14.33	5.877	157	0	1350
Nereid	5,513,400	222.64	360.16	157	0.75	170

Triton

Almost all our knowledge of the physical properties of Triton was learned from the data gathered by the *Voyager 2* probe's exploration. On August 25, 1989, after its fly-by 5,000 km (3,100 miles) from Neptune's cloud cover, *Voyager 2* approached Triton and logged its mass and radius. The density, twice that of water, reveals that it is mainly composed of ice and rocks. Triton, like Titan, has an atmosphere; both data from *Voyager 2*,

Above: Images of 1989 N1 and 1989 N2.

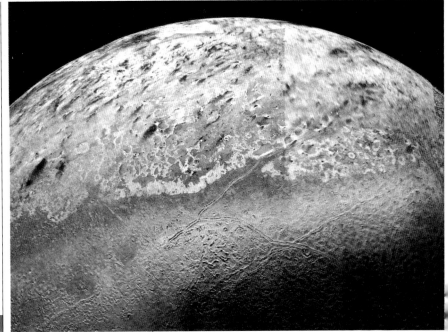

Above: The ice crust of Triton, furrowed by grooves and fractures; bright and darker zones alternate.

Triton, satellite of Uranus

Mass (g)	2.14×10^{25}
Mass (Moon = 1)	0.29
Mean density (g/cm³)	2.07
Mean density (Moon = 1)	0.62

and spectroscopic observations carried out from Earth indicate that its atmosphere consists of molecular nitrogen and methane, with traces of argon and carbon monoxide. The surface pressure measured by *Voyager 2* is fifteen-thousandths of Earth's atmosphere and the temperature is 38 K. Triton is therefore the coldest body found in the Solar System to date. Neptune's rapid rotation causes flattening of the poles of the planet and its equatorial bulge causes Triton's orbit to precess. This means that the plane of its orbit periodically changes its inclination in relation to Neptune's equatorial plane. As a result, at certain stages Triton exhibits a weak seasonal cycle, while at other times the satellite is so positioned that its seasons come to resemble those of Uranus.

Surface

When *Voyager 2* had its close encounter with this satellite, the Sun was high above Triton's south pole. The image of this region reveals bright zones, pinkish in color, which probably results from the white of the frozen nitrogen and the color of the methane, which turns red when exposed to solar radiation. There are very few craters. It would therefore seem that processes must have been at work that erased evidence of the numerous impacts that Triton must have undergone during the course of billions of years, while their traces persist in such numbers on many other bodies in the Solar System.

During the time when *Voyager 2* was observing Triton, it was summer at the south pole, and the ice was therefore slowly evaporating. The polar cap of Triton seems to be confined within a clear boundary. At lower latitudes, where the ice has melted, the terrain underneath was seen to be dark in color. Even within the polar cap, however, darker regions occur among the areas still covered with ice, and these are probably caused by volcanic activity or geysers. On the ice-free terrain, there is a clearly visible long fissure which forks in the shape of a "Y." Similar features have been observed also on the Jovian satellites Europa and Ganymede. These features were formed when a quantity of molten material flowed through the cracks in the surface, and then solidified, creating a sort of embankment along the crack. The presence of these formations shows that in the past the interior of Triton must have been in a molten state.

Surface Relief Features

In an image of the southwest margin can be seen three dark patches, irregular in shape and ranging between 100 and 200 km (60 and 120 miles) in size, with pale, bright edges. It seems that these are fairly recent structures which could be the result of volcanic activity, although this would be somewhat surprising given the fact that they occur in the region most liberally scattered with craters and therefore in terrain which formed in the most remote past. We know that the vast majority of impacts occurred during the first billion years of the Solar System's existence. Recently-formed terrains have erased and eroded the ancient craters. Another image taken by *Voyager 2* shows a walled plain that is almost perfectly circular in shape, 200 km (120 miles) in diameter, its surrounding rim of hills approximately 200 meters (600 feet) high. Within this plain, which is like a frozen lake in appearance, there is an impact crater, sited just off-center, which was evidently formed after the basin had filled with molten material. The crater is 13 km (8 miles) in

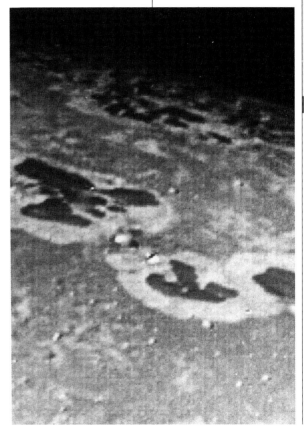

Below: A detail of Triton's surface. Three darkish patches, irregular in shape, are visible, varying from 100–200 km (60–120 miles) in length, surrounded by paler, bright material. These could be the results of fairly recent cryovolcanic activity.

Right: A photograph taken in front of the screens of Pasadena's Jet Propulsion Laboratory which was in charge of the Voyager 2 mission. This photograph captures the enthusiasm of the scientists when, on August 25, 1989, the actual appearance of Neptune was revealed to them by the cameras of the probe which had been launched from Earth 12 years earlier.

Below: A photo-montage reconstructing the amazing encounter between Voyager 2 and Neptune, the penultimate planet of the Solar System, at that time further away than Pluto which was traveling along the part of its orbit that comes closest to the Sun.

diameter and its highest point is 1 km (3,300 feet) above the lowest point. Apart from having many features in common with other satellites, Triton does have some unique features. Among these is the terrain near the border between the dark zone and the light zone, which is wrinkled, almost corrugated, bearing some resemblance to the skin of a melon and therefore aptly called "cantaloupe terrain." There are other frozen lakes which may have been formed by a form of ice volcanism. The fact that Triton is a celestial body that is still active is demonstrated by two images taken at an interval of 45 minutes with a detail resolution of 1 km (3,300 feet). In these an erupting geyser can clearly be seen: the column of this eruption is approximately 8 km (5 miles) high, exploding upwards from the surface, sweeping debris of dark particles along with the outrush, apparently producing smog in the satellite's atmosphere. As it rises, this column of gas produces a long, dark plume, which is wafted downwind and dragged out to a length of approximately 150 km (90 miles) by Triton's winds. Other, similar dark cloudy formations are driven northwards by the satellite's gentler winds and can be discerned in the more detailed images of the polar cap.

Activity on Triton

When *Voyager 2* transited behind Triton, the discovery of an ionosphere was a surprise for the scientists in charge of the mission. Examination of Triton's spectrum, carried out on Earth, confirmed that the satellite is a body that is subject to change. In fact, a spectrum of 1979 shows that the strongest emission took place in the bright-orange-red region, while another spectrum taken ten years later indicated a virtually constant emission in relation to the wavelength. In order to find an acceptable explanation for this, it must be supposed that the nitrogen and methane ice has covered a considerable area of the disk that is visible from Earth, whereas ten years earlier it was possible to glimpse part of the surface with its reddish hydrocarbon compounds. We have already mentioned that Triton's retrograde motion suggests the satellite was not part of Neptune's original system, but it must have been captured later by the planet's gravitational field. However, if this is the case, we would expect it to have an elliptical orbit, whereas Triton's orbit is circular. Probably, however, Neptune has caused gigantic tides on Triton, changing its orbit. Such tides would have heated it by friction and would have liquefied its interior, with resulting volcanic eruptions, as shown by the surface. This heating would also have created a dense atmosphere around Triton, because the frozen nitrogen, methane and carbon monoxide would have been transformed into gas and a certain quantity of these would have injected into the atmosphere by volcanic eruptions. Because of a greenhouse effect, the atmosphere would, in turn, have heated the surface, causing sublimation of more ice. The presence in the past of a thick atmosphere would explain, among other things, why the surface of Triton has so few craters on it and appears so smooth.

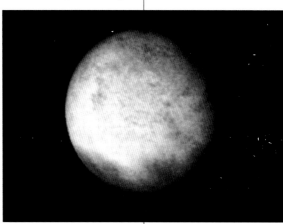

Below: An image of Nereid.

Bottom: An image of Triton, Neptune's largest moon, taken by Voyager 2 which passed in front of it at a distance of 40,000 km (25,000 miles). NASA's space probe discovered six new moons around Neptune, their diameters varying from 54 to 400 km (33 to 250 miles).

Nereid, Proteus, and Larissa

Voyager 2 succeeded in taking a photograph of Nereid from a distance of 4.7 million kilometers (2.9 million miles), during the last quarter phase but no detail can be made out in it. The largest of the newly-discovered satellites, Proteus, was however observed at a distance from which it was possible to show that its surface is thickly peppered with craters. Larissa, second in order of size of the satellites discovered by *Voyager 2*, was photographed and the image shows that its appearance is virtually identical to that of Proteus.

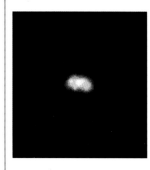

PLUTO ♇

Top: An image of Pluto from the Hubble space telescope, 1994.

Above: A reconstruction of Pluto and its satellite Charon. The proximity of the latter to its planet and relatively small disparity in size contributes to their definition as a double planet.

Physical characteristics

Pluto and its satellite Charon, the latter being only 19,640 km (12,200 miles) distant from its planet, describe a very elliptical orbit around the Sun: at aphelion they are 49 astronomical units away and are therefore the outermost bodies of our Solar System; at perihelion they are 29 astronomical units away and are consequently within Neptune's orbit, which then becomes the outermost planet of the Solar System.

With a radius of 1,142 km (710 miles), Pluto is not only the smallest of the planets, it is also smaller than many satellites, including the Moon and Jupiter's four Galilean planets as well as Titan and Triton. Pluto's single satellite, Charon, insofar as present knowledge goes, has a radius of 595 km (370 miles), approximately half that of Pluto. For this reason it is possible to describe these two bodies as "a double planet" rather than a planet-and-satellite system.

Charon's orbit around Pluto is perfectly circular, it has a radius of 19,640 km (12,200 miles) and is almost perpendicular to Pluto's orbital plane. This orbital plane is intersected twice by the Sun during the Plutonian year, which is equal to 248.77 terrestrial years: at such times, Pluto and Charon occult one another when observed from Earth. As a result, the system's light varies regularly with a period of 6.387 days.

Photometric variations (changes in the light intensity from the system as the bodies periodically occult each other) have enabled astronomers to obtain a considerable amount of data: apart from information regarding Charon's orbit, it has led them to discover that the density of this system is equivalent to 2.07 times that of water, similar to the density of Triton, and that the total mass of the system is equal to approximately 0.25% of the Earth's mass. The stellar occultation method is probably the most accurate when it comes to determining the dimensions of a planet or an asteroid and in revealing the presence of an atmosphere, if any. On June 9, 1988 Pluto occulted a star and this made it possible to confirm that a weak methane atmosphere exists on the planet. Charon also occulted a star, a few months after it had been discovered, but on that occasion no trace of any atmosphere was detected.

Satellites

The extreme closeness of Charon to Pluto explains why it was only discovered in 1978: telescopic images only allow for the presumption of Charon's presence, betrayed by a slight bulging in the disk of Pluto. Recent images obtained by the Hubble Space Telescope, which can distinguish details 10 times smaller than those discernible by terrestrially-based telescopes, have revealed the two small disks of these bodies, clearly separate. The Hubble telescope has also made it possible to measure Charon's mass, which is equivalent to one-twelfth that of Pluto.

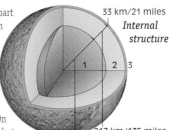

Internal structure

33 km/21 miles
217 km/135 miles
900 km/600 miles

1—Rocky core
2—Water-ice mantle
3—Crust of frozen methane

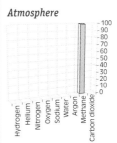

Atmosphere

Characteristics of Pluto

Mean distance from Sun (AU)	39.53	Mass (g)	1.29×10^{25}
Mean distance from Sun (10^6 km)	5319.52	Mass (Earth = 1)	0.002
Orbital period (days)	90,800	Equatorial radius (km)	1,142
Mean orbital velocity (km/s)	4.74	Equatorial radius (Earth = 1)	0.179
Orbital eccentricity	0.2482	Mean density (g/cm³)	2.05
Apparent mean diameter of Sun	49"	Mean density (Earth = 1)	0.37
Inclination of orbit to ecliptic (°)	17.148	Volume (Earth = 1)	0.006
Number of satellites	1	Ellipticity*	0.0

Equatorial surface gravity (m/s²)	1.29×10^{25}
Equatorial surface gravity (Earth = 1)	0.002
Equatorial escape velocity (km/s)	1.142
Sidereal rotation period at equator	6 days 9 h 17.6 min
Inclination of equator to orbit (°)	98.3

*Ellipticity is $(Re-Rp)/Re$, where Re and Rp are the planet's equatorial and polar radii, respectively.

Below: The photographs that led to the discovery of Pluto. The planet is the white point indicated by the arrows: the two images were taken on the different dates when observations were made and show the changes in the star's position. The photograph on the left was taken on February 23, 1930, the one on the right six days later, on February 29.

Bottom: A photograph taken from the U.S. Naval Observatory which led to the discovery of Charon: the satellite appears as a protuberance that bulges out slightly to the right of the upper limb.

A BELATED DISCOVERY

At the beginning of the twentieth century, the American millionaire and amateur astronomer, Percival Lowell, developed a passionate interest in the puzzling perturbations of the orbits of Uranus and Neptune. He also embarked upon photographic research, hunting for what was then widely considered a hypothetical planet, exploring the plane of the ecliptic.

From 1906–1916 thousands of photographic plates were examined, using very labor-intensive procedures. In 1915 it was suggested from study of perturbations that it was possible that a planet was present, with a mass equivalent to 6.5 terrestrial masses, sited at a distance 42 times as great as that between the Earth and the Sun, in the constellation of Gemini. Lowell died in 1916 and the task of continuing this research was subsequently entrusted to Clyde Tombaugh, who took over in 1929 using more up-to-date instruments.

From 1929 through January 1930 photographs were taken of the

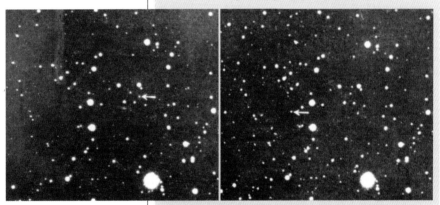

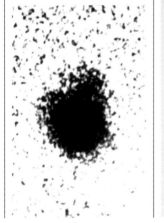

zones in the region of the sky near the ecliptic that was in opposition to the Sun when the apparent motion of the hypothetical planet was faster. The photographs revealed the presence of a planet in positions near to those theorized by Lowell, but of a size and brilliance that were far below his predictions and with an eccentric orbit. The research program was extended until 1943; meanwhile, technological progress meant that more accurate analyses could be carried out. This meant that it was possible to achieve a clearer explanation for the perturbations attributed to the planet that had been found and led to the realization that it had a smaller mass than Earth.

After the discovery of Pluto's satellite in 1978, the planet's mass was re-evaluated on the basis of reciprocal motions: this gave the result of 0.0026 terrestrial masses, a value that could not possibly have influenced the orbits of Uranus and Neptune. It must be acknowledged, however, that Pluto was finally located near to the point forecast by Lowell.

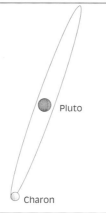

Orbit

What can account for Pluto's strange orbit, inclined by 17° to the plane of the ecliptic and so markedly elliptical? Is it because Pluto is accompanied by a satellite that is so large in relation to its paired planet? Why do Neptune's two largest satellites, Triton and Nereid have orbits that are so anomalous to those of its other satellites? All these questions have produced various hypotheses. Among others, one suggests that Pluto, Triton and Nereid could all be satellites of Neptune, sited on regular orbits. In a close encounter with a hypothetical body of planetary size, the orbits of these three satellites could have been strongly perturbed. Pluto, especially, would have split into two parts, settling into an orbit around the Sun. So far, however, this is still only a hypothesis which seeks to find out the origin of one of many mysteries of the Solar System.

Charon, satellite of Pluto

Mean distance from Pluto (km)	19,640
Distance in R_p	17.08
Orbital period (days)	6.387
Orbital inclination (°)	98.8
Orbital eccentricity	0
Radius (km)	595
Mass (g)	1.07×10^{24}
Mass (Moon = 1)	0.02
Mean density (g/cm³)	2.07
Mean density (Moon = 1)	0.60

Top, left: A comparison of size and inclination of rotational axes of Pluto and the Earth; top of page center: Charon's orbit.

Top, right: An image of the Pluto-Charon system taken from Earth and, above, an image from Hubble space telescope, showing Pluto and Charon clearly as two completely separate bodies.

THE MINOR BODIES

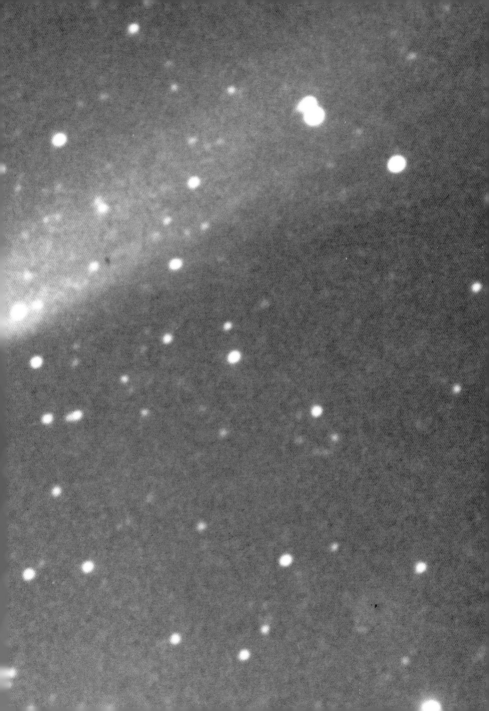

COMETS

Above: Comet West (1976 VI), photographed in March, 1976.

Facing page, top: A false-color image of the nucleus of Comet Halley taken by the Giotto probe.

Physical characteristics

A comet is a celestial body with a nucleus in a misty shroud or "coma," which has a long tail visible for a limited time. When the comet travels away from the Sun, the tail is the first element to disappear, followed by the nebulosity around its head: only the central part, the nucleus, then remains. The deduction that the nucleus was a small solid body was confirmed when the European probe *Giotto* succeeded in observing Comet Halley close up during its passage in 1986.

The nucleus measures approximately 10 km (6 miles) and is irregular in shape; it consists of ice, silicates and other volatile organic substances. A comet's nucleus differs from the asteroids in that it is richer in volatile material and usually has a far more elongated orbit. When the nucleus approaches perihelion and is subjected to the pressure exerted by ultraviolet solar radiation, material evaporates from the comet, forming a cloudy coma around it and a tail of dust is created which is the most spectacular cometary feature, while the ionized gas of the ultraviolet solar radiation forms a tail of gas which is distinguishable in color and shape from the dust tail. After reaching perihelion, the comet travels away and reverts to its insignificant appearance of a minute, dark body.

The number of new comets discovered every year suggests that the number of objects that will eventually turn into comets must be huge. The Dutch astrophysicist, Jan Oort followed this line of investigation and has proposed the existence of a large reservoir of cometary nuclei, their orbits having aphelions that would fall between 30,000 and 50,000 astronomical units. Planetary perturbations would cause them to venture into the inner Solar System, altering their orbits in such a way that their perihelia would be shifted from a distance further away from the Sun than Neptune.

Classification

Comets are classified according to their periods of revolution, which vary from the 3.3 years of Comet Encke to 1,000 years for the 1887 II comet, or the second comet that was discovered in 1887. In addition, a large number of comets have parabolic or hyperbolic orbits, that is, open orbits. After reaching perihelion, therefore, they disappear into interstellar space and it is difficult to find out whether this

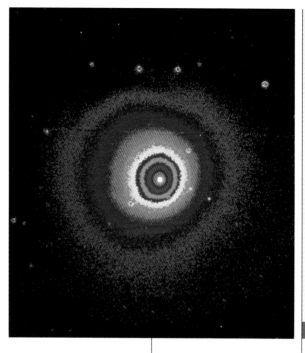

is due to an open orbit or to a very elongated elliptical orbit; orbits believed to be open could, instead, be elliptical with even longer periods of revolution, just as it is possible that open orbits can be modified by planetary perturbations. The first group of comets, described as short-period comets, have periods of between 3.3 and 20 years and orbits with properties similar to those of the planets, that is, inclined at relatively small angles to the ecliptic; only 14% have inclinations in excess of 20°. Nearly all move around the Sun in the same direction as the planets, that is, in an anti-clockwise direction as viewed from "above" the plane of the ecliptic. The aphelion of the great majority of these comets is situated in the vicinity of Jupiter's orbit.

Due to its short period, Comet Encke's passage to perihelion has been observed more than 50 times. Apart from being subject to perturbations caused by Jupiter, Mars, the Earth and Venus, its orbit exhibits additional variations. In

Below: On July 23, 1995 an unusually bright comet was sighted by two amateur astronomers, Alan Hale and Thomas Bopp, independently of one another. The comet, 1000 times brighter than Comet Halley has a diameter of approximately 40 km (25 miles).

Below, left: The orbits of the external planets, from Jupiter to Pluto and the orbit of Comet Halley. All the orbits of the terrestrial planets are contained within the white frame.

Below, right: The orbits of the terrestrial planets, from Mercury to Mars and the orbit of Comet Encke.

fact, solar radiation causes sublimation of the ice on that part of the nucleus facing towards the Sun, and this causes an action/reaction impulse similar to that which drives a rocket. Since the nucleus rotates, this can cause secular (slow) acceleration or deceleration in the orbital motion depending on the rotational direction of the nucleus.

Life of the comet

The effect of sublimation means that each time the comet travels to perihelion a part of the nucleus is transformed into the coma and tail, which

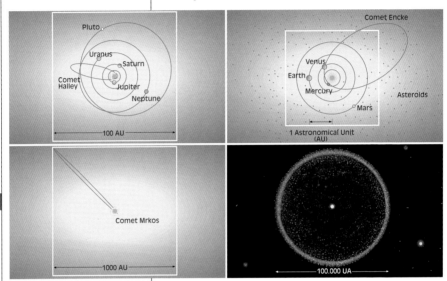

Above, left: The orbit of the Comet Mrkos which, at its maximum distance from the Sun, plunged into the Oort cloud; the latter probably begins at a distance of a thousand astronomical units, where an astronomical unit is equal to the distance between the Earth and the Sun.

Above, right: A diagrammatic illustration of the Oort cloud which is thought to extend for as far as 100,000 astronomical units—and further. The little dot in the center has a diameter of 1000 astronomical units and it is therefore over 20 times larger than the axis of Pluto's orbit.

will eventually disperse into interplanetary space. It is estimated that after about one hundred passages to perihelion, a short-period comet has completely evaporated. Consequently, the life of a comet belonging to this group is extremely short when compared with the age of the Solar System. It is probable that this is what finally happened to the Comet Biela, which had a period of 6.6 years, and various passages of which were observed during the period between 1772 and 1846, the latter being the year in which the nucleus was seen to split in two. In 1852 two separate comets appeared, 2 million km (1.2 million miles) apart, both very active. In 1866, the year in which it was forecast that they would make another appearance at a position favorable for terrestrial observers, the two comets failed to appear and nothing more was known of them.

The second group of comets are long-period comets, which have a life of more than 200 years. These have randomly distributed orbital inclinations and semi-axes greater than the orbital ellipses of between 10,000 and 100,000 astronomical units; they usually have larger nuclei than those of the short-period comets.

Left: An illustration showing the encounter between the European Space Agency's Giotto *space probe and Comet Halley, on March 14 1986, at a record minimum distance of 596 km (370 miles) from the nucleus of the comet. The probe crossed through the comet's coma, surviving the impact with its constituent particles.*

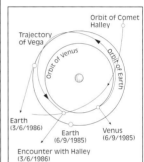

Above: The trajectory of the Russian Vega *probes which had their encounters with Halley on March 6 and 9, 1986.* Vega 1 *had been launched from Baykonur on December 15, 1984 and* Vega 2 *on December 21 that year. Before reaching their destination, these two probes had carried out fly-bys of Venus.*

A small number of comets also exists, described as intermediate-period comets, with periods ranging from 20 to 200 years: the most famous of these being unquestionably Comet Halley. A third of these comets orbit around the Sun with retrograde motion, with inclinations from 17° to 85°; in this respect they have more in common with long-period comets than short-period comets.

A group that has been observed mainly as a result of the Solar Maximum Mission (SMM), although some had already been spotted in the past, are the comets that pass very close to the Sun, almost skimming its surface or "sungrazing." From 1980 to the present day about ten of these sungrazers have been observed. This type of comet is at a very minimal distance from the Sun, ranging from 0.0055 and 0.067 astronomical units; when they are at perihelion, these comets come to within distances ranging from approximately only 800,000 to 10 million km (500,000 to 6 million miles) to the Sun. Some disintegrate at perihelion, others are broken up into fragments. It is possible that a certain number of comets are extrasolar in origin.

Below: A comparison of the abundance of certain atoms in the universe, in the interstellar grains and in the comets. Hydrogen (H) is by far the most abundant element; however, its abundance in the frozen interstellar dust and in comets is much lower than in the universe (stars and interstellar gas), since it is present in these objects only as a molecular compound, while hydrogen gas cannot be retained by bodies with such a small mass as comets, let alone the tiny grains. In the three charts: C=carbon, N=nitrogen, O=oxygen, S=sulfur, Si=silicon.

Coma and nucleus

The coma is formed by the frozen gases around and in the nucleus. When the nucleus approaches perihelion its outermost, volatile part is freed, producing a sort of atmosphere. The carbon monoxide and carbon dioxide, which are the most volatile substances, vaporize as far off as 10 astronomical units, which means beyond Saturn's orbit, but the ice only vaporizes at 3 astronomical units, in the asteroid belt between Mars and Jupiter. Because the mass of the nucleus is too small to hold onto the gas, this disperses into interplanetary space, while other material vaporizes, constantly renewing the cometary atmosphere. Vaporization occurs only on that side of the comet facing the Sun, given the great difference in temperature between the sunward side and dark side. The composition of the coma of Halley's Comet has been studied from close-up by various probes and, in particular, by the *Giotto* probe during the comet's passage to perihelion in 1986. If the quantity of water vapor present in the coma of the comet is taken to be 100, the other substances present in the greatest quantities are formaldehyde (4.5), carbon monoxide (5–10), carbon dioxide (3), methane (from 1 to 2) and other substances in smaller quantities.

The study of the spectrum of the coma can also provide important information as to the nature of the solid nucleus. When part of the nucleus vaporizes in the coma, it is possible to measure the presence and amount of elements, either in the form of molecules or as carbonaceous and silicon granules as impurities in the frozen gases which, as they

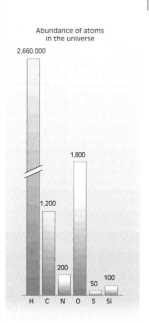
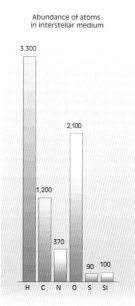
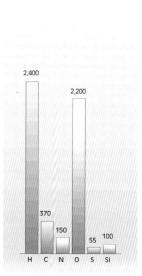

Left: The Pleiades constellation. To the bottom right of the photograph Comet Halley can be seen. This photograph was taken on November 16, 1985.

Below: Comet Mrkos (1956 V) photographed during the month of August, 1957. The images of the stars appear trailed, because the telescope follows the movement of the comet, which is far faster than the apparent movement of the stars due to the Earth's rotation.

vaporize, carry these along with them. The nucleus of Halley's comet, for example, when at a distance of 1 astronomical unit from the Sun, released 30 tons per second of gas and 24 tons per second of dust. Since the dust particles were microscopic granules with a diameter of a tenth of a micron, it is estimated that the nucleus freed 4,000 billion, billion granules per second. The dust is pushed radially outwards from the nucleus by the gases, while the dust travels much more slowly in a transverse direction.

If the outgassing from the nucleus is localized, jets are formed which can extend as far as 100,000 km (60,000 miles) or more. These jets, which escape in a straight line from the nucleus, are then progressively curved or "bent" by the pressure of solar radiation which accelerates the individual particles so that their motion is directed away from the Sun. The smaller the granules, the greater the acceleration; thus, in the dust tail of a comet a separation occurs between the dust particles according to their size.

The space missions towards Halley's Comet revealed the presence of molecules composed of carbon, hydrogen, oxygen and nitrogen, apart from the silicates already mentioned.

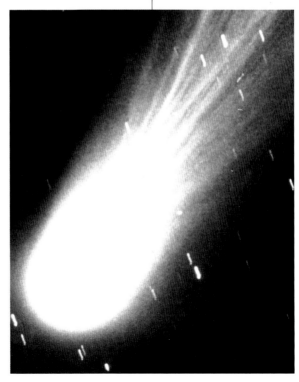

Right: An image of Comet Halley taken on February 22, 1986.

Below: The nucleus of Comet Halley taken by the Giotto probe's Halley Multicolor Camera.
1. Bright spots
2. Craters
3. Bright region
4. Mountain
5. Terminator

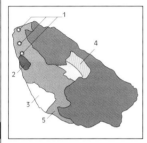

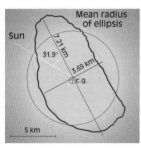

Above: A profile of the nucleus of Comet Halley (in red); the ellipsis is shown in yellow and this is a closer approximation of its real outline; violet denotes the center of gravity (c.g.) and the circumference which has as its diameter the mean of the axes of the ellipsis.

Temperature

The temperature of the coma when it is at a distance of 1 astronomical unit from the Sun is approximately 200 K, equivalent to 73°C (163°F) below zero, at the moment when the gas escapes from the nucleus. Subsequently, as the coma expands, it cools until it reaches approximately 20 K, the equivalent of 253°C (420°F) below zero. At distances ranging from 500-1,000 km (300-600 miles) from the nucleus, the very low density causes disassociation of the molecules and this process releases heat, taking the temperature of the gas up to 50-100 K.

The coma of hydrogen atoms freed from the water and OH molecules can reach dimensions of up to millions of kilometers. As an example, the ultraviolet OAO-2 space observatory observed the Tago-Sato-Kosaka comet in 1969 when its coma attained a radius of 30 million km (about 20 million miles).

Structure and composition of the nucleus

In 1950 the American astronomer, Fred. L. Whipple put forward a model for comet nuclei which has gone down in history as the "dirty snowball model." This consists of a structure of frozen ice, dust and gases, held together by mechanical and gravitational forces. Halley's Comet, of which the *Giotto* probe managed to take outstanding images before the camera was "blinded" by dust, confirmed Whipple's model. Its nucleus, shaped like a rather uneven potato, measures 8.2 x 8.4 x 16 km (5.1 x 5.2 x 9.9 miles). Radar observations and measurements of the brightness of the nuclei and their variations have enabled scientists to establish that the dimensions of most comets range between 1 and 10 km (0.6 to 6 miles) and are elongated ellipsoids with three axes.

We do not yet possess the means to take direct measurements of a cometary nucleus. From their dimensions and from an estimate of their density which, based on their composition is between 0.2 and 1.2 times the density of water, it can be inferred that a typical mass can measure from 1 billionth to 1 thousand billionth of the Earth's mass. The rotational periods of the nuclei estimated for some twenty comets range from four hours to seven days.

Despite the observations carried out by the space probes, where Halley's Comet is concerned, however, the value is uncertain, and is estimated to be between 2.2 or 7.4 days. Probably this is due to the fact that its triaxial shape produces a complicated rotation motion.

The European *Giotto* and the Russian *Vega* probes, have measured the albedo of the nucleus of Halley's comet, which is 0.027. In other words, it absorbs 97% of incident light; this is why it is almost black in color. Observations of the nuclei of various other comets reveal that their color can vary from differing shades of gray to dark red. These different shades depend on the varying relationships between surface dust and ice and on the varying composition of the surface. The nuclei are mainly constituted of water in frozen form, of refractory dust and organic compounds.

The presence of ice formed of water and from extremely volatile substances such as carbon monoxide and methane is clear proof that the nuclei were formed at very low temperatures, ranging from 15 to 30° K, far away from the Sun.

The Comet's Tail

Rather than referring to tail in the singular, it would be more accurate to talk of several tails, since each comet has at least two which are different in nature: there is the tail of dust and the tail of ionized gas or plasma, and their composition and evolution differ.

The dust tail is formed from a rarefied cloud of solid particles, expelled by vaporizing gas. The two forces act along an imaginary line connecting the comet to the Sun and are inversely proportional to the square of the distance from the Sun: we are talking of the force of gravitational attraction and of the force of radiation pressure, which repels. The smaller the particle, the stronger the latter force and this is what causes the separation of particles of different sizes: the larger, measured in millimeters, stay closer to the nucleus, while the particles measurable in microns or submicrons' disperse quickly into the tail, separating according to their

Above: A drawing of the nucleus of Halley's comet, as envisaged by an artist before it had been photographed by the Giotto probe.

Below, left: The structure of a comet. The solid part or nucleus has a mean diameter of approximately 10 km (6 miles). The coma can have a diameter ranging from 100,000 and 1 million km (60,000 and 600,000 miles). The tail can be as long as 1 astronomical unit (150 million km/93.2 million miles). In this diagram we see the two tails, the straight tail of ionized gas and the curved dust tail, pointing away from the Sun

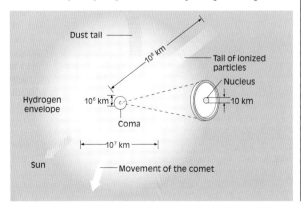

Below: An image of fragments of the comet Shoemaker-Levy 9, traveling towards Jupiter 660 million km (410 million miles) away from Earth.

Below: The "scars" left by some of the 21 fragments of the comet Shoemaker-Levy 9 which fell onto Jupiter between July 16 and 22, 1994. For the first time mankind could observe the spectacular collision between two bodies of the Solar System.

size and traveling away from the Sun. The length of the tail as seen from the Earth depends on the angle subtended by the tail of the comet relative to the line of sight. If this angle is large, we see the tail at its real extension. The progressive diminution of this angle increases the effect due to perspective and when the Earth crosses the orbital plane of the comet, visibility of the tail is minimal. Other factors influence visibility, among them the intrinsic brightness of the tail, together with the quantity of dust present in it and, of course, the greater or lesser degree of contrast with the brilliance of the sky.

"Anti-tails"

The effect of perspective also produces what are sometimes referred to as "anti-tails," or tails pointing in towards the Sun, which have occasionally been observed in certain comets. If large dust grains are present, that is if they can be measured in millimeters, they are subjected to a very weak radiation pressure and their trajectories are separated from those of the smaller particles which appear more strongly directed, pointed away from the Sun. The larger-sized particles can find themselves projected on the celestial sphere as if they were in front of the coma. Anti-tails are fairly rare, because they require very particular geometric conditions.

Although the dust tails do not usually display details, sometimes it is possible to see structures caused by episodes when a large quantity of dust is expelled from the nucleus, or even explosive phenomena occur, such as the break-up of the nucleus into two or more parts. Or shapes similar to spines (*aculei*—points) are observed, either pointing towards the Sun, or away from it. This effect is due to the distribution of the particles on ellipses around the nucleus: the further away they are from it, the smaller the size of the particles, because of the combined action of the forces of gravitation and of pressure from solar radiation. When the Earth is close to the orbital plane of the comet, the ellipses are seen edgewise-on and appear as pointed, sharply tapering jets.

Tails and plasma

The plasma tail is the result of the interaction of the comet with the solar wind. When the gas com-

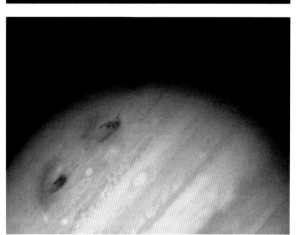

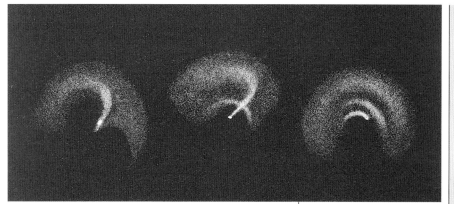

posed of neutral molecules expands outside the nucleus, it undergoes a process of ionization caused by solar ultraviolet radiation and forms an ionosphere around the comet which blocks the solar wind. The force lines of the Sun's magnetic field which the solar wind carries with it, wrap themselves around the comet and create a magnetic tail, composed of two lobes of opposing polarity. The plasma tail becomes visible because the molecules of carbon monoxide, ionized by the magnetic field and excited by solar radiation, emit radiation at a wavelength of 4,200 Å, corresponding to the color blue; this is why the tail has a bluish appearance. Additionally, the solar wind, which at a distance of 1 astronomical unit from the Sun, equivalent to the distance between the Earth and the Sun, which has a speed of approximately 400 km/s (250 m/s), accelerates the cometary molecules, and is, conversely, decelerated by them.

Study of the spectrum of the gassy tails indicates that ionized molecules of water, OHs and carbon dioxide are also present. The tails vary from a few tens of kilometers to 100 million km (60 million miles) in length, and their width can be a tenth or even less of their length. Unlike the dust tails, the plasma tails are straight and form an angle of approximately 5° pointing away from the Sun and away from the comet's movement.

Plasma tails have many features such as condensations, knots, and linear structures, called rays, which change from day to day. Among the most interesting and frequently-occurring phenomena are known as separations: the entire plasma tail detaches itself from the head and slides away in the opposite direction to that of the Sun, only to be replaced by another plasma tail.

The Oort Cloud and the Kuiper Belt

Although the known Solar System extends to beyond Neptune's orbit, at a distance of approximately 30 astronomical units, the Sun's gravitational field extends to at least 100,000 astronomical units—the equivalent of two light years. It is in this space that the remains of the

Above: The head and the coma of comet 1881 III in three drawings made on June 17, 27 and 28, 1881 respectively, based on observations carried out at the Geneva Observatory. These very accurate drawings clearly show the smallness of the comet's nucleus. It can be seen that the jet ejected on June 27 in the direction of the Sun then folds backwards towards the tail.

Below: The spectacle presented by the Donati's comet in 1858 in the sky above Harvard University.

cloud of gas and dust from which the Solar System condensed is to be found, as well as dust and gases that constitute the interstellar galactic medium.

In 1950, based on his observations of comet behavior, Jan Oort proposed that there was a spherical region at the edge of the Sun's gravitational field where comets resided. This distant region, now known as the Oort Cloud, is considered to be a source of long-period comets—as many as one trillion may originate there. Long-period comets tend to have highly elliptical, eccentric orbits. Indeed, what are believed to be open-orbit comets may actually be long-period comets with extended elliptical orbits.

The orbits of short-period comets, however, in addition to being of shorter duration, are also usually more predictable than those of long-period comets, so it is not surprising that their origins are different. Indeed, many appear to come from another area in the outer solar system known as the Kuiper Belt. This disk-shaped region lies beyond Neptune's orbit, about 30 to 100 astronomical units from the Sun and, in addition to acting as a reservoir for short-period comets, it also contains the planet Pluto.

Due to its distance from us and the small size of individual comets, the Oort Cloud has not yet been directly observed. However, in 1992, a galactic body 240 km (150 miles) wide was detected beyond Neptune, at a distance relative to that of the hypothesized Kuiper Belt, and it was soon confirmed that the Belt did exist. Since then, the study of the region along with the objects found in it (Kuiper Belt objects, or KBOs) has become an emerging area of research.

Origin and duration of the comets

A comet forms at temperatures lower than 25 K, as can be demonstrated by the presence of carbon monoxide, methane and molecular sulfur, all of which are extremely volatile at slightly higher temperatures. The comet probably passes the greater part of its life in the Oort cloud at temperatures of approximately 4 K. In this environment the comet

Below: The Sun is at the center of the Oort cloud: if the latter encounters a large interstellar cloud the gravitational perturbation produces a shower of comets inside the Solar System. These encounters can occur with a certain periodicity, because the Sun, in its motion around the galactic center shifts periodically slightly above and slightly below the plane of the galactic equator, and it is on this plane that the majority of the clouds of interstellar medium are to be found.

Right: The Sun is at the center of the Oort cloud. When a faint star (a red dwarf, given that these are by far the most numerous in the Galaxy and therefore also in the vicinity of the Sun), indicated by the red spot, enters the Oort cloud, the gravitational perturbation thus caused sends a shower of comets into the inside of the Solar System. A hypothesis has also been put forward, not yet confirmed by observations, that the Sun has a companion of a very small mass and with a very elongated orbit; its perihelion would fall within the Oort cloud. The theory suggests that every time this supposed body passes to perihelion, a shower of comets is produced and that this happens every 30 or more million years.

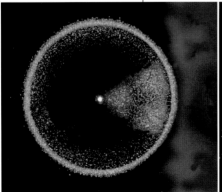
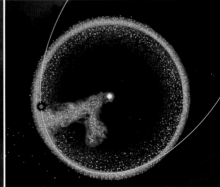

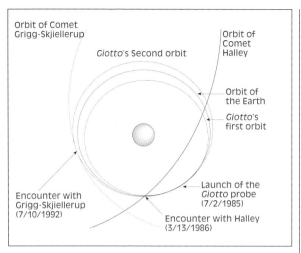

Left: The trajectory followed by the Giotto probe to its encounter with Comet Halley at a speed of 68.4 km/s (42½ miles per second). During this voyage two major orbital corrections were made and other, minor corrections, in response to data gathered by the Russian Vega space probes which had embarked upon their missions before this encounter.

Below: The Giotto probe during its assembly by British Aerospace, the lead manufacturer for this project. The probe had a diameter of 1.87 meters (6 feet 1 inch) and its external surface was covered with solar cells which delivered a 190-watt power supply; on board the probe carried a payload of 10 scientific instruments.

undergoes erosion processes caused by collisions with the dust of the interstellar medium and from cosmic rays. In the innermost regions of the Oort cloud, where the orbital velocities are greater and the density of objects is greater, it is probable that collisions occur with boulders and icy residues.

When the perihelion of the comet is nudged into the Solar System following perturbations caused by the outermost planets, or even by a star that is relatively close to the Sun, one of two things can happen: the comet could be expelled from the Oort cloud and will describe a hyperbolic orbit; or it can be captured within the Solar System on an elliptical orbit and live for a greater or lesser length of time depending on the duration of its orbital period and on whether it comes closer or not so close to the Sun at perihelion.

Short-period or intermediate-period comets are, however, likely to have lives lasting for a few thousand years, a negligible stretch of time in the context of the age of Solar System.

Above: Comet Halley's passage in 1066 was depicted in the famous Bayeux Tapestry. This work of art, an embroidery on linen measuring a total of 70 meters (240 feet) in length, tells the story of the Norman conquest of England. Tradition has it that the tapestry was embroidered by Queen Mathilde, the wife of William the Conqueror. (By kind permission of the City of Bayeux.)

Below: On the inner margin of a ninth-century manuscript, probably originating from Verona, Italy, and now housed in Padua's Antoniana library, there are drawings, albeit not conventional ones, of a comet with a double tail, which appeared not long after the death of Charlemagne.

THE RETURN OF THE COMETS

Comets are known for their unexpected disappearances and flaming tails and have had a powerful fascination for humankind since time immemorial. Aristotelian theory placed them in the sphere of the Earth, in the place were there could be generation and corruption, but they were soon credited with playing a part in the movement of the celestial sphere. During the Renaissance, astronomers discovered that comets do not have a discernible parallax and this indicated that they were very distant, way above the Moon, in the astral skies: from this stemmed the attempt to define their movements accurately. Kepler studied the motion of comets within the context of the Copernican system. Extrapolating the effects of terrestrial rotation, he established that the trajectories of the comets, from the moment of their appearance to when they disappeared in the brighter light of the Sun, appeared rectilinear. It was still believed that a comet traveling away from the Sun was different from one traveling towards it. Flamsteed was the first to suggest the theory that the comet of 1680–81 was one and the same. Newton studied the same comet, stating that it must have passed behind the Sun: using gravitational laws, he managed to plot its course geometrically, showing that it described a parabola or a very elongated ellipsis. Edmund Halley, after discovering the comet of 1682, subsequently named after him, calculated its previous passages, which had taken place in 1541 and 1607. He was able to calculate an orbital period of approximately 76 years, predicting its return in 1758, and when that year came, the comet punctually reappeared. This English astronomer's work proved to be a great step forward in the application of celestial mechanics.

GIOTTO

The return of Halley's Comet in 1986 played a part in stimulating further exploration of comets through projects launched in the United States, Soviet Union, Japan and Europe. The USICE satellite crossed the tail of the short-period Giacobini-Zinner comet and reached to within a distance of 28,100,000 km (17,500,000 miles) of Halley; the Japanese probes Sakisake and Suisei studied its dispersal and extensive interaction with the solar wind.

The Russian Vega 1 and 2 probes succeeded in obtaining very good images of the comet, which were also used to help improve the trajectory of Giotto which was constructed by the ESA. Giotto was launched by an Ariane 1 rocket on July 2, 1985 on the first European interplanetary mission; the probe was covered in solar cells capable of delivering a 190-watt power supply; it had a high-gain disk antenna and carried a scientific payload of 10 instruments, among these was a spectrometer, magnetometers and particle analyzers. It weighed 584 kg (128 lb) and was stabilized by means of a rotary movement of 15 turns per minute, reduced to 4 during the encounter. After two course corrections, Giotto passed by the nucleus of Halley's comet on May 14,1986, at a distance of only 596 km (370 miles), crossing the turbulent coma and sending 2,112 images back to earth, captured by its Ritchey-Chretien telescopic camera. The automatic control system, built by an Italian company, Laben, meant that it was possible to recover the probe after a temporary loss of control and contact, and it was used again on the following Giotto Extended Mission in 1988, which studied the Grigg-Skjiellerup comet. This comet has been eroded to a greater extent by the Sun than Halley's comet; this meant that crossing through its tail in 1992 was a marvelous chance to compare two comets which differ greatly from one another.

Below: The launch of the Ariane 1 rocket from French Guyana on July 2, 1985. The entire mission was coordinated from the European Space Agency's ESOC center in Damstadt, Germany.

ASTEROIDS

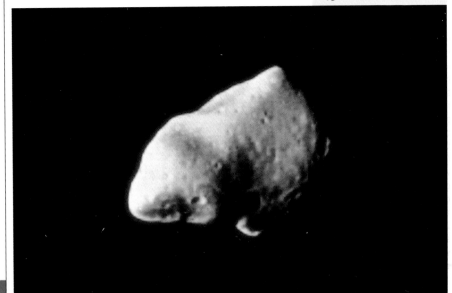

Above: The asteroid 951 Gaspra imaged by the Galileo probe on October 29, 1991 from a distance of 16,000 km (1,000 miles). This was the first "close-up" image of Gaspra ever taken.

Physical characteristics

The number of asteroids, or minor planets, of which the orbits are known with enough precision so that they can be easily plotted, number about 4,000; it is believed, however, that there are over 40,000 in existence that measure over 1 km (3,300 feet or 0.6 miles) in diameter. Nearly all the asteroids are found between 2.2 and 3.3 astronomical units and form the so-called "main belt" of the asteroids, while a few orbit nearer the Sun and a small number occur beyond the belt. In the asteroid belt the density of the orbits is not uniform: there are zones that are more crowded and empty zones, or nearly empty, where the periods of revolution around the Sun are simple fractions of that of Jupiter. These orbits are defined as being resonant with Jupiter and the zones in which they are found are called "the Kirkwood gaps" after their discoverer. The main ones occur between 2.06 to 3.28 astronomical units, with corresponding orbital periods that vary from one-quarter to one-half the orbital period of Jupiter. Other, more sparsely-populated zones occur at 2.96 and 3.03 astronomical units and correspond to orbital periods of $3/7$ and $4/9$ of that of Jupiter. Then there are three groups of asteroids sited outside the principal belt, at 3.97, 4.29 and 5.2 astronomical units with periods equivalent to $2/3$, one-quarter of Jupiter's period or coincident with it. These three groups are called "Hilda," "Thule" and "Trojan," respectively.

"Families"

It has been observed that those asteroids in which the semi-axes are greater than the orbital ellipses have values that are very close and also have similar orbits, exhibiting the same eccentricity and inclination on the plane of the ecliptic. This is where the term "family" comes from, in order to indicate that the members have common origin. It is important to recognize the members of a family because if they are fragments of a single body, it is possible to know immediately what the internal composition of one is by studying the others' remains. Statistical analysis of the frequency of families can give an indication of the break-up rate of asteroids inside the belt. However, it is not easy to recognize the members of a family, because of the instrusion of numerous other, extraneous bodies which are also present in the same belt. Much remains to be learned about those asteroids with a diameter of less than 40 km (25 miles) belonging to the main belt. The same applies to asteroids with diameters of over 1 km (3,300 feet or 0.6 miles) and which come close to Earth, and there are thought to be at least a thousand of them; only 10% of these have been cataloged to date. A very effective technique is that of sending radar pulses and picking up the echo; from this information can be gathered, including the orbital movement and the surface features. Some asteroids are very efficient reflectors of radar waves, and from this it can be inferred that their composition must be metal-rich.

Below: A diagrammatic illustration of the asteroids' orbits. Most of these are in the main belt, between Jupiter and Mars. The orbits are almost circular and only slightly inclined on the ecliptic. Another group moves along very elongated orbits and are very sharply inclined to the plane of the ecliptic.

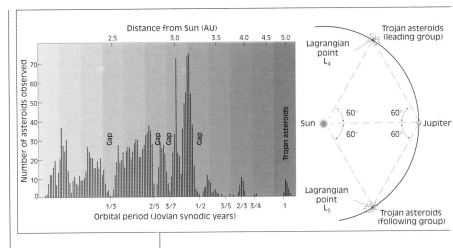

Above, left: The vertical scale of the graph shows the number of cataloged asteroids; on the horizontal scale are the distances from the Sun in astronomical units, and, along the bottom of the graph, their orbital periods: the latter are expressed taking Jupiter's synodic period (one year and one month) as the unit, this being the time that Jupiter takes to return to the same region of the sky as seen by a terrestrial observer (also in motion, because he or she is part of the Earth's revolutionary motion). It can be seen that there are gaps, empty, or almost empty, of asteroids corresponding with fractions of Jupiter's synodic period equal to ⅓, ⅔, ⅗, ½. This typical distribution is due to gravitational perturbations caused by repeated alignments with Jupiter and the Sun.

Above, right: The Trojan asteroid family is grouped within what are known as the Lagrangian points L4 and L5. Each of the two groups occurs at the vertex of an equilateral triangle, of which the other two vertices are the Sun and Jupiter. These two Lagrangian points have the property whereby bodies that occur there maintain a constant position in relation to the other two vertices of the triangle, i.e. the Sun and Jupiter.

Mass, density, rotation

Little is known about the asteroids' masses and densities but from the few cases studied, it would seem that they possess similar densities to rocks and meteorites. The greater part of mass is concentrated in the few large bodies. In fact, the total mass of all the asteroids is equal to about five-hundredths of that of the Moon and is barely three times greater than that of Ceres, the largest asteroid. Since they have an uneven shape, their capacity for reflecting the Sun is anything but uniform. During their rotational movement, the asteroids have different parts of their surfaces turned towards the Earth, and their luminosity therefore appears to change; by observing the variations in light as a function of time, the so-called "light curve," it is possible to find out the rotation period. Some asteroids rotate in approximately two hours, but the majority have periods of eight or nine hours, while many others rotate very slowly with periods of several weeks. From the aspect and width of the curve of light it is also possible to extrapolate data on the shape of the asteroid and about the inclination of its rotational axis. If the asteroid is spherical with a fairly uniform surface, the light curve will exhibit a regular progression without extreme maxima and minima. If however, the shape is ellipsoidal, the light curve exhibits two maxima and two minima, depending on whether the asteroid has its two surfaces of maximum area or minimum area facing a terrestrial observer. The longer the asteroid, the greater the difference between maxima and minima. In addition, the duration variation also depends on the inclination of the rotational axis. This is at its maximum when the terrestrial observer is viewing the equator, but it is completely flat, that is without any variation, if the asteroid has one of its poles turned towards the observer.

Composition

The composition of asteroids can be determined by examining their spectra. Those with a low albedo (very dark) are similar to the carbonaceous chondrites and have almost certainly the same chemical composition; they are denoted with the letter C. Others, also with a low albedo but more reddish in color, are indicated by the letter P. Then there are others, more red, that are given the letter D and these are the most common among the most distant asteroids, outside the main belt. Another group, indicated with the letter S, has a higher albedo and is more frequent in the innermost belt. The asteroids denoted by the letter A are composed of olivine, like the Earth's mantle. The letter M signifies those asteroids rich in metals, as is suggested by the high reflectivity of the radio waves sent by radar. The letter E stands for asteroids with a high albedo and which are, presumably, composed of enstatite, an iron-poor silicate. The presence of absorption lines provides further information on the nature of asteroids. In general, it can be said that the same range of chemical composition present among meteorites also applies to asteroids.

Chemical variations

The most interesting aspect of the various chemical compositions of the asteroids resides in the fact that they vary according to their distance from the Sun. Type E, highly reflective, predominates in the innermost edge of the main belt; type S, with a fairly high albedo and a reddish color, occupies approximately one-third of the innermost part of the belt, while type C occupies the other two-thirds. Type P predominates in the zone outside the asteroid belt, but inside the orbit of Jupiter; type D is typical of the family of the Trojans sited on Jupiter's orbit.

The hypothesis has been put forward that this regular variation in composition depending on distance from the Sun could reflect the variation of the composition of the nebula cloud from which the Solar System originally condensed and of which the asteroids are themselves an example. Such an hypothesis would, however, run the risk of over-simplification: many classes of asteroids show different degrees of thermal alteration. It is therefore more plausible that their composition changes depending on how far they are from the Sun because these bodies have been subjected to a different degree of heating. These rocky or metallic bodies whose orbits come close to that of the Earth, are interesting in

Above: The various shapes typical of asteroids with a diameter in excess of 200 km (120 miles). Real shapes are shown here and their sizes are to scale: for purposes of comparison, the limb of Mars is shown (left).

Below: A photograph of two asteroids (arrowed). The telescope rotates around an axis parallel to the terrestrial rotational axis with the same angular speed as the Earth, in order to keep the stars fixed within the telescope's field of vision. Because of their closeness to the Earth, asteroids have an appreciable movement of their own, and they shift during observation, forming an elongated image.

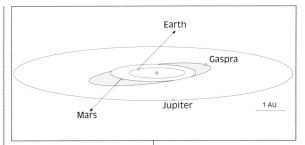

Above: Gaspra's orbit, shown with that of Jupiter, Mars and the Earth.

Below: In this photograph, taken in 1882 by Max Wolf in Heidelberg, an asteroid has left an elongated trace or "trail" to which the arrows point. This was one of the first asteroids discovered by means of photography and it was called Svea.

many ways. Most of the meteorites that fall to Earth come from the asteroid belt. Thus meteorites bring to the Earth samples of material from regions beyond the Earth-Moon system. Moreover, the impact of these objects with the terrestrial planets has had, and can still have, significant effects on their surface.

Orbital features

Asteroids have been classified as to their orbital features, into three groups that take their name from three typical asteroids: 1862 Apollo, 1221 Amor and 2062 Aten. The objects of the Apollo group have, generally speaking, orbits that are almost completely external to those of the Earth, but with a perihelion that is less external to the Earth's orbit which is at 1.017 astronomical units. Their orbits can cross that of the Earth. The asteroids of the Amor group have larger semi-axes than the orbital ellipse in excess of 1 astronomical unit and perihelion at distances greater than the terrestrial aphelion, in general ranging from 1.017 and 1.3 astronomical units. The orbits of the Amor group are all external to the terrestrial orbit. The division into these two classes

Left: An image showing the encounter between Galileo and Gaspra.

Below: A map drawn up from observations made by IRAS (Infra-red Astronomical Satellite). The positions of 1811 asteroids are shown (the colder ones in red, the hotter ones in black) and, as a point of reference, the orbits of Earth, Mars and Jupiter are also shown.

Bottom: A detail of the zone occupied by the Trojan asteroids (for greater precision those which precede Jupiter are known as the Greeks, those which follow, Trojans). Jupiter, the Sun and the two groups of asteroids occupy the vertices of two equilateral triangles, having the Jupiter-Sun side in common.

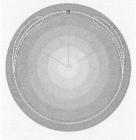

is not clearly defined because perturbations can transform an object belonging to the Apollo group into one belonging to the Amor group and vice versa.

Objects belonging to the Aten group, however, have orbits which are almost completely within those of the Earth, and they intersect the latter's orbit when they are at aphelion. The larger semi-axis of the orbital ellipsis is less than 1 astronomical unit. These asteroids have short lives and are continually being replaced by members of the other two groups.

Among the asteroids grazing the Earth, 1862 Apollo was the first to be discovered in 1932, while 2062 Aten was only discovered in 1976. Today over 100 such objects are known, of which only seven belong to the Aten group, and all have been discovered during the last 15 years, as a result of a systematic observation campaign carried out by many astronomers all over the world. From the statistics based on the rate of discoveries, it is estimated that approximately 500 asteroids belong to the Apollo group; 1,500 to the Amor group; and 100 to the Aten group, which are brighter than the magnitude 18, meaning they are 63,000 times weaker than a star at the limit of visibility to the naked eye. From this value of luminosity it has been deduced that the maximum dimensions for objects with low albedo are up to 1.7 km (5,600 feet or 1.05 miles), and 0.9 km (3,000 feet or 0.6 miles) for those which are more reflective.

Size

Today we know the size and the albedo of approximately a quarter of the asteroids that come close to the Earth. Three objects of the Amor group have diameters in excess of 10 km (6 miles). They are: 1036 Ganymede, 38.5 km (24 miles); 433 Eros, 22 km (14 miles) and 3553 1983SA, 18.7 km (11.6 miles) in diameter. The smallest asteroid of which the diameter is known is 1915 Quetzalcoatl, which is barely 500 m (1,600 feet) long. Their albedo falls mainly within the parameters of 0.01 and 0.53, with mean values of 0.20 and is therefore quite high. The population of the

Above: The asteroid Ida with its little moon, Dactyl taken in visible light by NASA's Galileo probe with its CCD camera on August 28, 1993. The dimensions of Ida are approximately 58 x 25 km (36 x 15½ miles), those of Dactyl 1.5 km (.93 miles/4,921 feet). Ida belongs to a family of over 200 asteroids called Koronis, originating from the fragmentation of a single, larger body.

main asteroid belt is dominated by dark objects and these are four times as numerous as those with high albedo. It must be remembered, however, that bright objects are more easily spotted than the darker ones and it is therefore probable that among the smaller objects, the number with low albedo has been underestimated.

"Life span" of the asteroids

The life span of those asteroids near to the Earth is destined to be far shorter than that of the Solar System. Their orbits are subject to such perturbations that in the course of 10 or 100 million years they will fall on one of the large planets or they will be expelled from the Solar System. Those that are most susceptible to strong perturbations are asteroids with very elliptical orbits and low inclination. The fact that the asteroid population near to Earth does not die out suggests that it is continually replenished by new members, coming either from the asteroid belt or by capturing short-period comets. To cite one example of the latter case: observations of the nucleus of Halley's comet by the *Giotto* probe suggest that when the surface ice evaporates after a sufficient number of passages to perihelion, a body of carbonaceous and silicate material will remain, making it indistinguishable from an asteroid.

A problem that closely concerns the inhabitants of Earth is the probability of an impact with an asteroid, which could have disastrous consequences for the planet, or for a large part of it. It must be remembered, however, that although the orbits of the minor planets of the Apollo and Aten groups intersect that of the Earth, they are in general tilted at considerable angles, by as much as 60° in relation to the plane of the ecliptic. As a general rule, therefore, at that moment when the orbits cross one another, the asteroid is far above or far below the plane of the ecliptic. Therefore the probability of collision is low and, given the fact that the number of large asteroids is very small, the probability of disastrous impacts becomes even lower.

Gaspra

When the *Galileo* probe was passing through the asteroid belt in October 1991, it acquired a number of images of the asteroid Gaspra, from a distance of 1,600 km (1,000 miles). The surface is peppered with craters: on the surface turned towards the camera some 40 were seen, apart from numerous grooves and fractures, there was evidence of many impacts.

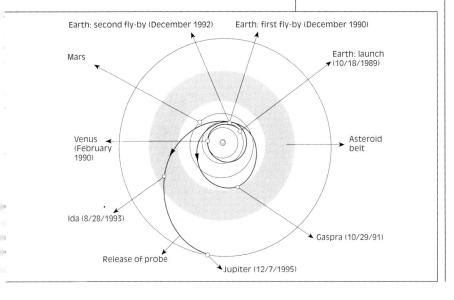

Below: The trajectory of NASA's Galileo probe which, during its journey towards Jupiter, encountered first the asteroid Gaspra on October 29, 1991 and then, on August 28, 1993 the asteroid Ida.

THE MISSING PLANET

The discovery of Uranus in 1781, at a distance that almost corresponded with that predicted by Titius-Bode's law, underlined the gap that existed between Mars and Jupiter and led Bode himself to undertake research into a hypothetical planet that could exist in that region. In 1801, a cleric, Father Piazzi, discovered the presence of a celestial body at a distance from the Sun 2.77 times greater than that existing between the Sun and the Earth and very close to the value of 2.8 suggested by Bode's law; this was called Ceres. Attempting to find it again after its temporary disappearance, the German astronomer Heinrich Olbers found another asteroid, called Pallas, with the same orbital characteristics as Ceres. Olbers hypothesized that the two bodies that had been discovered were fragments of a larger body which had shattered, and encouraged the search for further fragments, with the result that by 1895 four hundred of these were already known to exist. However, it was thought that these asteroids would not account for the body of a planet, until at the end of the nineteenth century when it became possible to carry out direct computation of their diameters, which turned out to be far greater than those suggested by the photometric method. Other important discoveries were made by Daniel Kirkwood, who noted that the distribution of the asteroids was not haphazard, but showed gaps where Jupiter's perturbations acted in a periodic way; then came the discovery of the Apollo asteroids, which move along orbits that intersect the Earth's orbit; that of the Lagrangian points, where some asteroids gather at points 60° in front of and 60° behind Jupiter in its orbit, and have the same period of revolution around the Sun. The advent of astronomical photography has since made the search for asteroids a far more fruitful one, enabling scientists to catalog them.

Above: Three images of Vesta obtained in 1987 at the Steward Observatory in Arizona using the "speckle interferometry" method. The irregular appearance of the asteroid is obvious and it seems to have rotated between the various exposures, taken at 15-minute intervals. At the time of these observations, the asteroid had a diameter of 0.5 arc seconds.

Below: A print celebrating the discovery of Ceres. The instrument shown on the right is Ramsden's altazimuthal circle, bought in 1789 for the Palermo Observatory and which was used to carry out the measurements that led to the discovery of Ceres.

Gaspra is a small asteroid; the side facing the probe is 16 km (10 miles) long. It is elongated in shape and very irregular. It orbits around the Sun at a mean distance of 331 million km (206 million miles), approximately 2.2 astronomical units. Its rotational period is 7 hours 2 minutes. Its albedo is approximately 0.20 and it is fairly uniform in color. False-color images, which exaggerate even the smallest details, show bluish regions, indicating an excess of olivine, while other, more reddish areas, show where that mineral is absent. It is not clear as to what type of rock underlies the regolith.

Gaspra is an example of a typical asteroid belonging to the main belt. Based on spectroscopic observations it can be classified as a type S asteroid, composed of silicates (i.e. stony material) and metals. Its pockmarked face, bearing the scars of encounters and collisions, suggests that it its surface is less than 1 billion years old; to be more precise, from the state of the visible craters, it can be estimated that the age of this asteroid could range from 300 to 500 million years. When the *Galileo* probe transited near the asteroid at a speed of 8 km/s (5 miles/s), it was feared that it might encounter dust or even tiny satellites of Gaspra, which might damage the probe's cameras. But the dust gauge on board Galileo did not register any impact with particles larger than those produced in cigarette smoke. This, too, was predictable, because such a small body does not have sufficient gravitational attraction to enable it to hold on to dust or satellites.

Above: The asteroid Gaspra, imaged on October 29, 1991 by the Galileo *probe. Gaspra is not a primitive body, but comes from the fragmentation of an asteroid that must have been larger. Its dimensions measure 19 x 12 x 11 km (11.8 x 7.45 x 6.84 miles).*

Origin of the asteroids

It is thought that asteroids may represent the original state in which matter existed 4.5 billion years ago: therefore, these fragments of primordial rock are also called "genesis rocks" and this intrinsic identity provides us with evidence of the past history of the Solar System and forms the most interesting facet of these small, apparently insignificant, celestial bodies.

The most widely accepted theory concerning asteroids is that Jupiter's presence originally prevented the formation of an hypothetical planet. It is thought that the only changes that the asteroids have undergone are those due to mutual collision and that the majority of them have not undergone thermal processes because of their small mass. The great number of members of the asteroid family therefore provides us with a

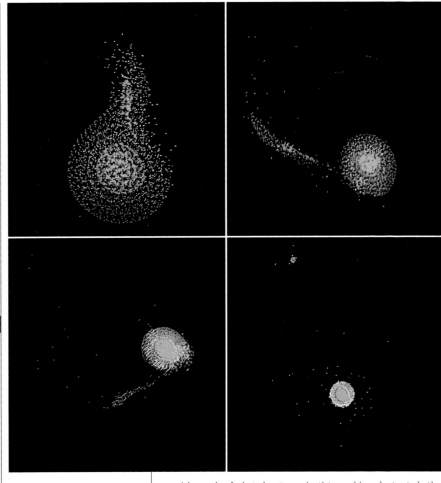

Above: A computer simulation of the effect on the Earth of an impact with a body of the size of Mars and with a mass one-tenth of that of the Earth, traveling at a speed of 12 km/s (7.5 miles/s). The nucleus is shown in blue, the mantle in bright orange. An enormous plume of vaporized rocks is produced which could give rise to a family of asteroids. Some scientists think that most of the rubble or detritus condensed to form a single celestial body, the Moon.

very rich sample of what planetary scientists need in order to study the origin and evolution of the Solar System.

Transneptunian bodies

In 1992 certain small bodies were discovered beyond the orbit of Neptune. Similar to comets, they were called "Kuiper belt objects." By late 2000, over 300 of these had already been identified but, according to astronomers' estimates, this number could increase to over one million. The average size is approximately 100 km (60 miles), but the majority are smaller, in the order of a few tens of kilometers. The transneptunians are presumably icy objects which are similar to comets. Most of them have moved little from their original position when the Solar System was formed, while others, subject to migration due to gravitational influxes, become short-period comets, with the aphelion of their orbit located beyond Jupiter. The existence of these bodies had been suggested by the Dutch-born American astronomer, Gerard Kuiper and others.

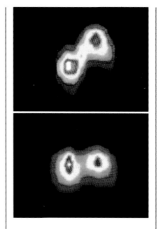

THE SEARCH FOR EROS

After NASA had completed the first phase of direct exploration of the first asteroids by means of the Galileo *probe, the Agency devoted an entire mission to the investigation of the asteroid 433-Eros. This mission was one of the small, inexpensive ones that fell under the heading of the "Discovery" program and it was planned that the NEAR (Near Earth Asteroid Rendevous) probe should go into orbit around the asteroid, carrying a payload of six instruments, including a laser altimeter to very accurately determine the shape of the asteroid. NEAR was launched on February 17, 1996 and in June 1997 made a fly-by of the type C asteroid, 253 Mathilde, sending back a great deal of images and scientific data. Then, after correcting a temporary fault, the probe went into orbit around Eros on February 14, 2000 and studied it for one year, descending to lower altitudes at intervals in order to carry out more detailed reconnaissance of the surface and eventually landing on the asteroid. NEAR has contributed to our knowledge of asteroids and also to the study of possible collision risks between these bodies and the Earth.*

Above: In August 1989 the asteroid 1989 PB which was approaching Earth was observed by radar echo technique, using the large Arecibo radiotelescope. The images were obtained from a distance of 5.5 million kilometers (3.4 million miles). This is a double asteroid, which could be produced when two bodies more or less the same size, orbiting close to one another, attract each other gravitationally.

Below: This drawing shows NEAR approaching an asteroid. During the course of its mission, this probe has gathered data and high-resolution images of the asteroids Mathilde and Eros.

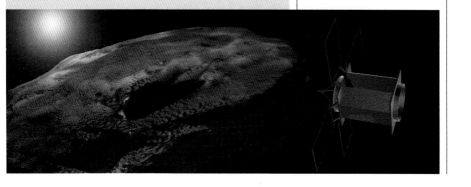

METEORITES

Above: A meteor radiant in the constellation of Leo. A meteor shower, shown here by arrows indicating the path across the terrestrial atmosphere, seems to have originated from a particular point in the celestial sphere, in this case the constellation of Leo.

Facing page, above: A Meteor crater in Arizona that is 1,200 meters (4,000 feet) in diameter and 180 meters (600 feet) deep, this is one of the largest, most recent and the best preserved craters of this type on Earth. The meteoroid that caused it was probably traveling at a sharply inclined angle and its fragments now lie under the edge of the rim. Despite this, the crater has a very symmetrical appearance.

Physical characterisitcs

During the course of its orbit, the Earth travels while immersed in the interplanetary medium, and in this are scattered, apart from neutral atoms and ionized particles from the Sun, many solid bodies, whose dimensions vary from many meters to a few microns. The surfaces of many of these bodies, penetrating the terrestrial atmosphere, undergo sublimation, that is, they pass directly from a solid to a gassy state, while some, larger bodies, undergo only partial sublimation. They fall to the ground, and these are called meteorites. Those which are completely sublimated before they reach the ground are called meteors. Finally, come the micrometeorites, those dust particles which because of their microscopic mass, manage to float gently down to the ground, intact. Both meteorites and meteors ionize the atmosphere by attrition, leaving a bright, very short-lived trail, usually referred to as a "falling star." Thousands of tons of micrometeorites fall onto the Earth each day. Although this quantity may appear huge, during the 4.5 billion years of the Earth's life, the fall of cosmic dust only accounts for an increase in the terrestrial mass of the order of a ten-millionth of its actual value.

Classification

Studies of the composition of meteorites investigate the concentration of chemical elements, isotopic composition, and their mineralogical composition. These are studies that are fundamental to an understanding of the nature of meteorites and to their parent bodies. Their isotopic composition enables scientists to find out how old they are.

Meteorites are classified according to their composition into three main groups: stones (aerolites), those mainly composed of silicates; stony-irons (or siderolites) composed in equal measure of silicates and iron; irons (siderites), composed mainly of iron and nickel. Aerolites are further classified as to whether they are chondritic or achondritic.

Chondrites are rocky meteorites composed of silicates containing iron and magnesium, as well as sulfur-bearing and metallic minerals. The majority of chondrites contain chondrules, which are small glass spheres a few millimeters in diameter which form as a result of the rapid crystallization of the molten material. Carbonaceous chondrites also contain agglomerates, but these are irregular in shape and of sizes ranging from a few millimeters to a few centimeters, with inclusions of minerals which only form at very high temperatures.

Achondrites are stony meteorites that resemble terrestrial igneous rocks, and they are thought to have resulted from large-scale melting and slow crystallization that occurred in the celestial body. Among the chondrites, the carbonaceous type is considered the best indicator of the Solar System's primordial chemical composition: a conclusion that follows comparison between the abundance of elements in carbonaceous chondrites and that of the Sun, when studied through spectroscopic analysis.

Analysis of isotopic composition

The study of the isotopic composition of meteorites helps scientists to establish the type of parent body, comparing the isotopic relationships of stable isotopes, and to find out its age, comparing the ratio of stable isotopes to unstable isotopes. For example, the ratio between oxygen isotopes of atomic weight 17, 18 and 16, which is by far the most plentiful isotope, indicates which meteorites have a "parent" in common and which belong to different families. Some achondrites found in the Antarctic are thought to be of lunar origin because their composition and, in particular, the three isotopic ratios

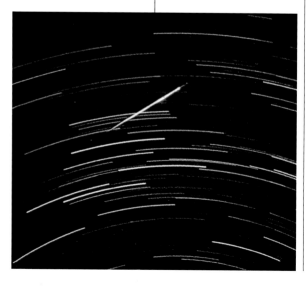

Below: The bright trail blazed by a meteorite. The luminous arcs are the traces or trails left by the stars due to the effect of the apparent motion of the celestial sphere from east to west but are due in reality to the rotation of the Earth from west to east.

of oxygen, are equal to those of certain rocks brought back to Earth by the astronauts of the Apollo lunar missions. A group of achondrites probably arrived from Mars, because the gas trapped in the minerals has a composition similar to that of the Martian atmosphere analyzed by the *Viking* probes. Other achondrites come from asteroids, as shown by their reflective properties in visible light and in infrared.

Chronology of the Solar System

In order to find out how long ago the minerals forming a meteorite crystallized, the ratio of radioactive nuclei to their stable product is measured. Examples of useful ratios for such measurements are those between potassium 40, of which the half-life is about 1.2 billion years, and its stable product, argon 40; or, the decay of uranium to thorium and to lead, or of rubidium 87 to strontium 87.

A number of anomalies have also been discovered, in the ratios of stable isotopes, which could not have been present in the primordial solar nebula. Such anomalies would seem to indicate an extrasolar origin

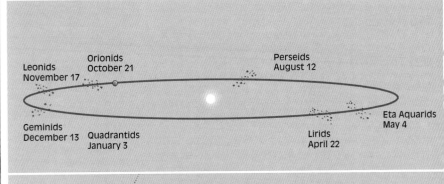

Below: In the diagram the Earth's orbit is shown and the points at which it intersects the orbits of old comets. The meteor showers are called after their radiant: for instance, on August 12 the Perseids will occur (the famous "tears of San Lorenzo").

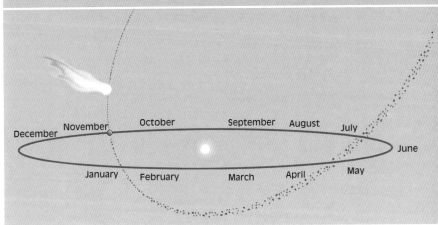

Above: In rare cases the debris left by a comet in its orbit tends to intersect Earth's orbit. When this happens, there is a shower of meteors.

of galactic material, enriched by stars in a very advanced state of evolution, which have synthesized these elements in their interiors.

It was believed that the Solar System originated from material at high temperature, produced by the collapse of the primordial nebula, which would have caused all the pre-existing, solid material to vaporize, wiping out every trace of chemical evolution that had come about under the influence of preceding generations of stars.

However, in some meteorites, among the most primitive known to us, isotopes of neon and oxygen have been discovered in percentages such that a solar origin has to be excluded and to allow for the origin being that of supernova stars or giant red stars and, in each case, stars far larger than the Sun, which had reached the end of their evolution.

Organic Substances

In the carbonaceous chondrites nitrogenous compounds have been found, present in living organisms and considered to be the building blocks from which the nucleic acids DNA and RNA are formed, and very complex organic compounds, such as amino acids, present in all known living things on Earth. Although there is a lingering doubt that some of these compounds were not originally contained in the meteorite, but are derived from contamination by the terrestrial environment, it would seem that, in some cases at least, this contamination can be excluded. A prime example is the Murchison meteorite which exploded on September 28, 1969 over Murchison in Australia, and which consists of different organic compounds from those found on Earth, with very different carbon isotopic ratios. Organic extracts from terrestrial sediments have in fact a ratio between carbon 12 and carbon 13 of approximately 40, while in the Murchison meteorite this ratio is approximately 20. If the organic compounds found in meteorites are of extraterrestrial origin and not due to contamination, new avenues of exploration open up to trace the formation of these compounds, from which, on Earth, the biochemical structures of the earliest life forms evolved.

Top: The Clovis aerolite remained underground for hundreds of years before being discovered.

Above: The Waingaromi siderite, belonging to a common sub-group of siderites, known as octahedrites.

Below: The Harriman siderite. The visible dark circle is an inclusion (a block of material in the matrix) consisting of the minerals troilite or ferrous sulfide.

Meteors and meteor showers

Particles of a few millimeters in diameter that penetrate into the atmosphere sublimate completely before they reach the ground. The atoms of these particles, on colliding with the atmospheric atoms and molecules, are ionized, and emit light. Those commonly referred to as "falling or shooting stars" are, in fact, the bright trails left by these particles during the process of vaporization. Two types of meteor can be identified: showers of meteors, which seem to

Right: A drawing of the fragmentation of a meteor, which was observed on October 19, 1863 by Julius Schmidt, director of the Athens Observatory and a well-known discoverer of comets.

Below: Samples of microtektites (microsopic glass formed by a large impact).

Left: A microtektite of 240 microns in diameter, with strange star-shaped craters; right, a microtektite from the Ivory Coast, enlarged 190 times.

come from a small region of the sky, and the sporadic meteors, distributed more or less randomly across the celestial sphere. The former, which can be observed at certain times of the year, are much more numerous and there can be as many as 50 of them or more per hour; the latter on average can be observed in numbers varying from one to five or six an hour.

Meteor showers

Certain types of meteor showers, such as the Quadrantids, the Perseids and the Geminids, are very regular and arrive on time above the Earth every year on the same date and with showers that are almost equally heavy. This means that the meteoric particles are distributed in an approximately even manner along their orbit: as a result, wherever the Earth intersects such an orbit, it encounters about the same number of particles. Other showers such as the Lirids, the Leonids and the Giacobinids are responsible for far heavier showers but only at intervals

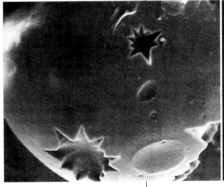
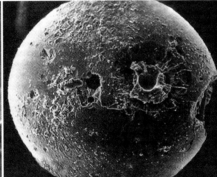

of several years and this means that the particles are grouped only at certain point of their orbit. Every so often, a shower disappears, or as the years pass the number of meteors from that shower becomes lighter. It has been observed that some of these showers follow the same orbit as a comet and, although in other cases no cometary orbit coincides with that of the showers, the most promising hypothesis as to the origin of the showers is that they are residues left by comets during their passage to perihelion and by comets that have already disintegrated. This would also explain why some meteor showers grow smaller as time goes by, because the cometary residue is gradually destroyed, both in encounters with the terrestrial atmosphere and through the effect of perturbations.

Sporadic Meteorites

The origin of sporadic meteorites is more uncertain. They could be residues of the nebula from which the Solar System formed, but they

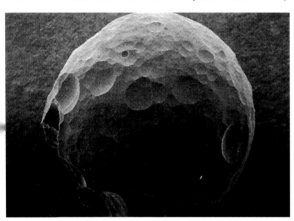

could also be dust coming from interstellar space. This makes it important to distinguish the type of orbit that could be open, that is, either hyperbolic or elliptical. This can be predicted on the basis of the speed of the meteor in question, bearing in mind that to this must be added or subtracted the Earth's orbital speed, which is 30 km/s (18.6 miles/s). If their speed in relation to the Sun is greater than 42 km/s (26 miles/s), which is the escape velocity of the Solar System at the distance between the Earth and the Sun, sporadic meteors will only deviate but not be captured inside the Solar System. If, however, this speed is less than 42 km/s (26 miles/s), they will follow elliptical orbits around the Sun, in common with all the other bodies in the Solar System. The first situation obtains if the comet comes towards the Earth, the second when it follows it. Therefore, we need to find out whether sporadic meteors exist with speeds in excess of 72 km/s (45 miles/s) in the first case, or of 12 km/s

Above: A microtektite peppered with craters, its appearance reminiscent of a microscopic Mercury.

Below: The zodiacal light, in a photograph taken in 1961 from a stratospheric balloon. This phenomenon is simply sunlight reflected by the dust scattered along the ecliptic, and it is probably connected with asteroids and meteorites.

Right: Examples of meteorites, illustrated in a book from the early twentieth century. Some samples, shown in this section, are of considerable complexity.

Below: Examples of meteorites belonging to various classes: rocky meteorites or aerolite (on left: Saint Mary's County) and stony-iron or siderolites (center and right: Krasnoiarsk and Mount Padbury). A section is shown of the St. Mary's County meteorite photographed with a polarizing microscope. The various colors indicate different materials. Siderolites are the least numerous class of meteorites. Krasnoiarsk was the first to be discovered belonging to this group. Crystals of olivine sparkle between the iron-nickel alloy matrix that surrounds them. Stony granules containing silicates (feldspar and pyroxene) are scattered throughout the Mount Padbury meteorite.

(7.5 miles/s) in the second. Terrestrial attraction, moreover, increases this speed by 1 to 5 km/s (0.6 to 3 miles). Out of many thousands of meteors, only 0.3% could have hyperbolic orbits.

Finally, it seems possible that sporadic meteors also form part of the Solar System. Collisions between sporadic particles and meteoric showers are frequent and can produce craterization on the granules themselves; microscopic observation of micrometeorites shows their surfaces to be pitted with microcraters.

Left: The Willamette Meteorite, discovered in the state of Oregon in 1902, weighing 16 tons.

THE TEARS OF SAN LORENZO

The smallest meteorites that appear in the sky evaporate in the highest layers of the atmosphere and their traces, called "falling or shooting stars," are only visible at night. The study of their dynamic has only started in relatively recent times. During the night of November 12 and 13, 1833, the Boston mathematician, Denison Olmsted, noticed that the radiant of the shower of shooting or falling stars he was witnessing shifted as time went by, keeping its position with regard to the celestial sphere, localized in the constellation of Leo. Hence the fact that these meteorites were called Leonids. Other showers could be relied upon to fall at a specific time of year, among which were the Perseids or "tears of San Lorenzo." While studying the latter in 1866, Schiaparelli realized why they made such regular appearances: on August 10 the Earth crosses the orbit of a comet and the Earth plows through a trail of meteors.

Above: This popular illustration shows a shower of shooting stars on the night of November 11, 1833. The depiction is not too far-fetched, when it is borne in mind that a likely estimate would allow for approximately 500 shooting stars a minute.

Left: The Leonids recorded during the nights of November 13 and 14, 1866 at Greenwich. The thickest lines indicate the luminous trail. It can be seen that the greater number of meteors seem to come from a single point in the sky called the "radiant."

APPENDICES

Aberration of starlight
The apparent displacement of a star due to the orbital movement of the Earth round the Sun. It can be observed by the combined effect of the finite speed of light (approx. 300,000 km/s/186,282 miles/s) and the Earth's orbital velocity (approx. 28 km/s/17miles/s).

Absolute magnitude
The magnitude (a measure of brightness) that a star would have if it could be observed from a distance of 10 parsecs.

Albedo
The reflecting power of a planet or other non-luminous body, expressed as the ratio of incident light received by a spherical body and the amount of light reflected by the same body.

Altitude
The angular distance of a celestial body above the horizon, ranging from 0° at the horizon to 90° at the zenith.

Ångström (Å = 0.0000000001 meter)
A unit of length equal to 10 to the power of negative 10 meters, used mainly as a measurement of the wavelength of light but also for other electromagnetic vibrations.

Angular momentum
A physical measurement expressing the "rotational momentum" of a body.

Angular velocity
A measurement expressing the rate of rotation of a body.

Aphelion
The orbital position of a celestial body in solar orbit that is farthest from the Sun.

Apogee
The point in the Moon's or an artificial satellite's orbit of the Earth at which it is furthest from the Earth

Apparent magnitude
A measurement which makes it possible to classify the stars according to the intensity of light (luminosity) received (i.e. index of a star's brightness relative to that of the other stars), as seen from a single location (i.e. Earth).

Astronomical unit (A.U.)
A unit of measurement equivalent to the mean distance between the Earth and the Sun, equal to 150 million km (93 million miles).

Azimuth
The angle between the vertical plane passing through a star and the meridian plane of the place of observation. The horizontal direction of a celestial point from a terrestrial point expressed as the angular distance from a reference direction.

Brightness, luminosity
The degree of intensity of light, emitted by a celestial body.

Celestial equator
The projection of the terrestrial equator onto the celestial sphere.

Celestial sphere
An imaginary sphere with an arbitrary radius, having as its center the Earth's center and its equator coincident with Earth's equatorial plane and upon which the celestial bodies are projected.

Conjunction
Two celestial bodies are said to be in conjunction when they have the same longitude. The Moon is called new when it is in conjunction with the Sun. For the inner planets, conjunction is said to be inferior when the planet transits between the Earth and the Sun, and superior when the planet transits across the far side of the Sun, furthest from the Earth.

Convection
One of the ways in which heat is transported in a fluid. The warmer "cells" of fluid, of lower density, tend to rise to the surface, while the colder ones tend to fall.

Cosmic rays
Elementary particles and atomic nuclei that move through space and bombard the terrestrial atmosphere from every direction.

Culmination
The transit of a celestial body on the observer's meridian; this occurs twice a day and is described as superior when the body reaches the greatest distance above the horizon, and inferior when the height above or below the horizon is minimal.

Declination
The angular distance on a celestial sphere measured from the celestial equator. It varies from 0° to 90° for a point in the northern hemisphere and increases towards the celestial pole; declinations are expressed as negative for points in the southern hemisphere and range from 0–90°.

Density
The ratio between the mass and the volume of a body (i.e. the mass of a substance per unit volume).

Direct (or prograde) motion
The movement of a celestial body around the Sun in the same sense as the Earth: that is, in an anti-clockwise direction for an observer situated at the North Pole.

Doppler effect
The apparent variation of the wavelength of light emanating from a moving source in motion in relation to the observer. If the source is moving away there is an increase in the wavelength and a shift towards red of the spectral bands (red-shift); if the source is approaching, there is a reduction in wavelength and a shift towards blue of the spectral bands (blue-shift).

Eclipse
The disappearance or reduction in visibility of a celestial body caused by its passage inside the shadow cast by another body.

Ecliptic
The apparent yearly trajectory of the Sun between the stars: it is defined as the maximum circle on the celestial sphere which represents the intersection of the terrestrial orbital plane with the sphere itself.

Electromagnetic spectrum
The distribution of energy emitted by a source in the form of electromagnetic waves as a function of the wavelength. Conventionally, this is subdivided into radio waves, microwaves, infrared, visible, ultraviolet, X and gamma, in decreasing order of wavelength. Each chemical element exhibits emission lines that are characteristic of it, and it is therefore possible to deduce the chemical composition of gases from their emission spectra.

Equinox
The two points of the celestial sphere in which the ecliptic intersects the celestial equator; they are therefore the nodes of the ecliptic. For the ascending node the Sun transits around March 21: this point is also called the spring (or vernal) equinox. For the descending node the Sun transits around September 21: this point is called the autumnal equinox. At the time of both equinoxes the length of the day is exactly the same as that of the night for every place on Earth.

Escape velocity
The minimum velocity necessary for an object to escape the force of gravitational attraction of a celestial body.

Gamma point
One of the two points of intersection between the ecliptic and the celestial equator and at which the Sun transits at the spring equinox.

Gravitational force
The force of reciprocal attraction which is exerted between all material bodies, directly proportional to the product of their masses and inversely proportional to the square of the distance between them.

Gravity assist
The exploitation of the gravitational energy of the planets by interplanetary probes when passing nearby, which enable them to alter their trajectory as required.

H alpha line (hydrogen alpha line)
The most prominent spectral line of the Balmer series: a series of spectral lines of hydrogen which fall into the visible region of the spectrum.

Hertzsprung-Russell (H-R) diagram
A graph that correlates the temperature of stars with their absolute magnitude.

Ion
An atom which, having lost or acquired one or more electrons, has a positive or negative electrical charge.

Kepler's laws
The famous laws that govern the movement of the planets around the Sun. The first law states that the orbits of the planets are ellipses whose common focus is occupied by the Sun; the second law states that the planetary orbits sweep out equal areas in equal times; the third law establishes that the squares of the periods of revolution of the planets are proportional to the cubes of their respective mean distances from the Sun.

Latitude
The angular distance measured along the meridian of a geographic locality starting from the Equator.

Light year
A unit of distance equal to the distance traveled by light in one year, the speed of light being 299,792 km/s (186,282 miles/s), a light year equals 9.46×10^{12} km.

Local Meridian
The meridian passing through the place where the observer is (i.e. observing site).

Longitude
The angular distance measured along the Equator from the Greenwich meridian of a given geographic locality.

Lunar month
There are two lunar months defined; the sidereal lunar month and the time taken by the Moon to complete a revolution around the Earth reckoned according to the fixed stars and which is equal to 27 days 7 hours 43 minutes; the synodical month (lunation) is the interval between two successive new moons and is equal to 29 days 12 hours 44 minutes.

Magnetic declination
An angle, at a given point on the Earth, between the plane of the magnetic meridian and that of the geographic meridian.

Magnetic field
The region surrounding a magnetized body in which such a body exerts a magnetic force; this entire region is affected by the strength of a magnetic force.

Momentum
The product of the mass and the velocity of a body; the momentum provides a measure of a body's inertia.

Nebulae
Celestial objects constituted by masses of gases mixed with dust particles. Their luminosity is due either to diffused light (reflection) from nearby stars, or to light emitted by gas excited by ultraviolet radiation from nearby hot stars. Dark nebula, conversely, are aggregates of cold and opaque matter that hide the star within from sight and therefore look like dark regions that do not contain stars.

Node
The point of intersection of the orbit of a celestial body with the ecliptic. If a body crosses the ecliptic from south to north the node is described as ascending; from north to south, descending.

Occultation
The covering up of one celestial body by another of greater apparent diameter.

Opposition
Two celestial bodies are said to be in opposition when their longitude differs by 180°. The opposition of a planet with the Sun occurs when such a planet is seen from the Earth in the opposite direction to the Sun.

Parallax
The apparent shift of a body when observed from two different directions or observing sites.

Perigee
The point of minimum distance in the orbit of a celestial body (the Moon) or artificial satellite that rotates around the Earth.

Perihelion
The point of minimum distance to the Sun in the orbit of a celestial body that rotates around the Sun.

Phases
The apparent changes in the shape of the Moon (and of the inferior planets) due to the different relative positions of the Earth, Sun and Moon (or planet).

Plasma
Gas composed of atoms that have lost one or more electrons, and of free electrons taken as a whole, it is therefore electrically neutral, since the positive charges and negative charges are present in equal numbers.

Precession
The apparent slow movement of the celestial poles due to the wobble of Earth's rotational axis. This involves a corresponding movement of the celestial equator and therefore of the equinoxes.

Pressure
The ratio between the perpendicularly exerted force on a surface element and the area of the element itself.

Red giant
A star that has exhausted its core fuel of hydrogen gas and has started on the phase of helium combustion. This triggers an expansion with an increase in radius by 100 times or more and a diminution of the surface temperature of the star, which therefore appears reddish in color.

Retrograde motion
The movement of a celestial body around the Sun (planet, comet, satellite) in a sense opposite to that of the Earth.

Roche limit
The distance from the center of a planet within which a satellite would be destroyed by the tidal forces exerted by that planet. The Roche limit lies at approximately 2.4 times the planetary radius assuming that the satellite has the same density as the planet.

Solstice
The days in which the Sun reaches its maximum distance from the equatorial plane. On June 21, the Sun reaches its maximum positive declination, +23°.5, (i.e. northernmost point); this is the Summer Solstice; on December 21, when the Sun reaches the maximum negative declination −23°.5, (i.e. southernmost point), this is the Winter Solstice.

Spectral line
An emission or absorption line in a spectrum: the former is a luminous segment which stands out against a black background; the latter is a thin, dark line which crosses a bright background (see spectrograph).

Spectroheliograph
An instrument that can analyze the surface of the Sun, by photographing it in particular wavelengths, (i.e. one wavelength only at a time).

T-tauri stars
Young irregularly variable stars which represent the last stages in the formation of a star, when the protostar is progressing towards the triggering of the first nuclear reactions within its own interior.

White dwarf
Very small, very dense stars that have consumed all their sources of nuclear fuel and are in a very advanced state of evolution.

Year
The time taken for the Earth to go once around the Sun. In the Gregorian calendar its duration is 365.25 days but from an astronomical point of view a year is defined as:
1) the sidereal year which has 365.25636 mean solar days (i.e. the time taken by the Sun to complete its path along the ecliptic in relation to the fixed stars) and,
2) the tropical year of 365.24220 mean solar days (i.e. the interval of time between successive passages of the Sun through the mean equinox).

Zenith
The point of the celestial sphere placed vertically to the observer (overhead point).

INDEX

Numbers in bold refer to photographs, illustrations and caption information.

Apollo 11, 111–12, **113**
asteroids, 224–35
   life span, 230–31
   orbits, **225**, 228–29
   origins, 233–34
   physical characteristics, 224, 226–29, 230
aurorae, **96**, 97, **98**

Callisto, 142, **143**, **144**, 147–48
Charon, 204, 205, **206**, 207
comets, 210–23
   classification, 211–12
   origin and life span, 212–13, 220–21
   physical characteristics, 210–11
   structure, 214–19

Deimos, **126**, 127, 128
Discovery, **44**, 86
Doppler effect, 32–33

Earth, 78–98
   atmosphere, 78, 82–85
   climate, 89, 97–98
   gravitational field, 80–81
   internal structure, **79**, 92–94
   magnetic field, 94–95, 97
   physical characteristics, 78–80
   surface, **83**, **84**, **85**, **88**, 89–92
Europa, 142, **143**, **144**, 146–47

faculae, 28, 30, 46

Galileo, 151, **229**, 231, 235
Ganymede, 142, **143**, **144**, 147
Giotto, 210, **213**, 214, 216, 217, **221**, 223
greenhouse effect, 64–65

Halley's Comet, 210, **211**, 213, 214, 215, 216, 217, 222
Hubble Space Telescope, **15**, 160, 205

Io, 142, **143**, **144**, 145–46

Jupiter, 132–51
   formation of, 18, 22
   Great Red Spot, 136–37, 148, 195
   internal structure, **133**, 137–38
   magnetic field, 138–40
   physical characteristics, 132, 133, 141
   rotation, 133–34
   satellites, 142–48
   study of, 149, 151
   surface, 134–35

Kuiper Belt, 219–20

Magellan probe, **68**, 69
Mariner probes, **60**, 61, 68, 129
Mars, 114–29
 atmosphere, 116–17
 climate, 117
 exploration of, 129
 physical characteristics, 114–16
 satellites, **126**, 127–28
 surface, 117–20, 122, 124–25
Mercury, 56–61
 atmosphere, 57
 internal structure, **57**, 58
 physical characteristics, 56–57
 temperature, 60
Moon, 99–113
 atmosphere, 107–108
 exploration of, 111–12
 internal structure, **100**, 105, 107
 magnetism, 107
 origin and evolution, 108–110
 physical characteristics, 99–101
 seismology, 103–105
 surface, **99**, 101–103, **105**, **106**, **107**

nebulae, **12**, 16–18, 21
Neptune, 188–203
 atmosphere, 189–90
 Great Dark Spot, **190**, 192, 195
 magnetic field, 195–96
 physical characteristics, 188–89
 rings, 196–98
 satellites, **191**, **198**, 199–203
 temperature, 194–95

Oort Cloud, **212**, 219–20, 221

Pathfinder, 129
Phobos, **126**, 127
Pioneer probes, 68, 69, 149, 151, 163
planets, 56, 78, 73–74
 formation of, 17–18, 21–22
 See also names of individual planets
Pluto, 204–207
 atmosphere, **205**
 internal structure, **205**
 physical characteristics, 204–205
 satellite, 204, 205, **206**, 207

Saturn, 152–75
 atmosphere, **153**, 154–55
 formation of, 18
 internal structure, **153**, 161
 magnetic field, 163–65
 physical characteristics, 152–53

 rings, **161**, **162**, **163**, **164**, 165–69
 satellites, 164, 166, 169–75
 spots, 157, 159–60
 study of, 149, 158
 temperature, 154, 155, 156
solar corona, **36**, 37–41, **50**, **52**, **53**
solar flares, 36–37
solar prominences, **26**, **32**, **34**, **35**, 36, **46**, **47**
solar spectrum analysis, 31–32
Solar System, 12–25
 life span, 13, 23
 formation of, 16–18, 24–25
 and formation of planets, 17–18, 21–22
 Sun and, 13–14
solar wind, 41–43, 46, 74, 75, **76**
stars, 14–16
Sun, 26–53
 atmosphere, 33–34
 internal structure, **27**, 34–36, **42**
 life cycle, 14, **22**, 23, 47, 50–52
 physical characteristics, 26–27
 radio emissions, 46–47
 and Solar System, 13–14
 surface, 28–31
sunspots, 28–31, **38**, 48–49

Titius-Bode law, 12
transneptunian bodies, 234–35
Triton, 199–203

Ulysses, 44–45
Uranus, 176–87
 atmosphere, **177**, 178–79
 internal structure, **177**, 179–80
 magnetic field, 180–81
 physical characteristics, 176–78
 rings, 181–83
 satellites, **177**, **180**, 183–86

Vega probes, **213**, 217, 223
Venera probes, 68, **69**
Venus, 62–77
 atmosphere, 63–65
 cloud cover, 77
 internal structure, **63**, 72
 magnetic fields, 73–76
 physical characteristics, 62–63
 surface, 66–67, 70–72
 temperature, 65–66
Viking probes, 115, 117, 119, 129
Voyager probes, 156, 163, 167, 187, 199, 200, 201, 203

BIBLIOGRAPHY

Stephen P. Marau (editor),
The Astronomy and
Astrophysics Encyclopedia,
Van Nostrand Reinhold,
New York 1992

Williams J. Kaufman III,
Discovering the Universe,
Freeman and Co., New York
1990

Geoffrey Briggs and
Frederic Taylor,
Atlante Cambridge dei pianeti,
Zanichelli, Bologna 1989

Roman Smoluchowki,
Il sistema solare,
Zanichelli Bologna 1989

Planetary Exploration
Through Year 2000,
SSEC Nasa 1988

Andrew Wilson,
Solar System Log,
Jane's, London 1987

D. Hart,
The Encyclopedia of
Soviet Spacecraft,
Exeter Books 1987

Reinhard, Battrick,
Space Mission to
Halley's Comet,
ESA 1986

Carl Sagan and Ann Druyan,
Comet,
Random House, New York 1985

Yeats and others,
Galileo: Exploration of
Jupiter's System,
Nasa 1985

Kendrick Frazier,
Il Sistema Solare,
Mondadori, Milan 1985

Oran W. Nicks,
Far Travelers,
Nasa 1985

E.C. Ezel, L.N. Ezel,
On Mars, Exploration of the
Red Planet 1958–1978,
Nasa 1984

Giovanni Caprara,
Il libro dei voli spaziali,
Vallardi, Milan 1984

Margherita Hack,
L'universo violento della
radioastronomia,
EST Mondadori, Milan 1983

Fimel, Colin, Burges,
Pioneer Venus,
Nasa 1983

Wenzel, Marsden, Battrick,
The International Solar
Polar Mission,
Esa 1983

D. Morrison,
Voyages to Saturn,
Nasa 1982

Soviet Space Programs
1971–75 and 1976–80,
U.S. GPO

Fimmel, Van Allen, Burges,
Pioneer: First to Jupiter,
Saturn and Beyond,
Nasa 1980

Dunne, Burges,
The Voyage of Mariner 10,
Nasa 1978

R. Cargill Hall,
Lunar Impact,
Nasa 1977

V.DP. Glouchko,
Encyclopédie soviétique
de l'astronautique mondiale,
Mir 1971

Giorgio Abetti,
La stella Sole,
Boringhieri, Turin 1966

http://pds.jpl.nasa.gov/planets
A site called Welcome to the Planets featuring a comprehensive
collection of images from NASA's Planetary Exploration Program.

http://mpfwww.jpl.nasa.gov
The site of NASA's Mars Exploration Program containing the latest
news, information and multimedia images.

http://cass.jsc.nasa.gov/moon.html
A site by the Lunar and Planetary Institute featuring all you need to
know about the moon and its exploration by astronauts.

http://www.earth.nasa.gov
The homepage of NASA's Earth Science Enterprise: a project dedicated
to understanding the Earth and the effects of natural and
human-induced changes on the global environment.

http://spaceflight.nasa.gov/station
A site containing news, images and information from the astronauts
currently working on the ISS (International Space Station).

http://seds.lpl.arizona.edu/billa/tnp/
A site called The Nine Planets which features a fascinating multimedia
tour of the solar system.

http://saturn.jpl.nasa.gov/index.cfm
Full coverage of the Cassini space probe, which is currently on its way
to explore Saturn and its moon Titan.

http://www.jpl.nasa.gov/galileo
All the latest news and images from the Galileo space probe during its
current journey to Jupiter.

http://encke.jpl.nasa.gov
A site containing information and images of comets and meteors as
supplied by NASA's Jet Propulsion Laboratory.

http://nssdc.gsfc.nasa.gov/planetary/giotto.html
Images and information about the Giotto space mission and its
studies of the comets Halley and Grigg-Skjellerup.

http://www.nationalgeographic.com/solarsystem
A 3-D virtual reality tour of the sun and planets, including
extraterrestrial weather patterns throughout the solar system.

http://www.seti.org
The SETI (Search for Extra-Terrestrial Intelligence) site. The project's
mission is to search for evidence of extraterrestrial life in the universe.

http://space.jpl.nasa.gov
A solar system simulator for students created by the Jet Propulsion
Laboratory at NASA.

http://solar-center.stanford.edu
A site that provides students with a collection of on-line educational
activities about the sun.

http://sohowww.nascom.nasa.gov
Images and data obtained by the SOHO (Solar and Heliospheric
Observatory) space probe.

http://www.bbso.njit.edu
A site containing studies and images of the sun as observed from the
Big Bear Solar Observatory in California.

PHOTOGRAPHS:

Anglo-Australian Telescope Board: 12, 13 (above)
W. Alvarez Collection: 90
Yves Langevin Collection: 125
Norbert Rosing Collection: 96–97
B.P. Glass, University of Delaware: 240–241
Caltech, Mount Palomar: 28
Centro Documentazione Mondadori, Milan: 78, 89 (below, left)
E. Petit: 34 (left)
ESA: 211
Gruppo Astronomico Tradatese, Tradate: 210, 215
Marshall Space Flight Center Institute: 50
IRAS-NASA: 16
Lunar & Planetary Laboratory, University of Arizona: 140
Millimeter Wave Astronomy Laboratory, Maryland: 159 (below)
Mount Wilson Observatory: 31
Harvard College Observatory: 38
High Altitude Observatory, University of Colorado: 41, 39 (below)
Lowell Observatory: 116 (left), 166 (left)
Sacramento Peak Observatory: 30 (above)
U.S. Naval Observatory: 23 (center), 39 (above), 166 (below)
Yerkes Observatory: 227 (below)
Smithsonian Institution: 239, 242 (center)
Space Research Institute, Moscow: 127
Pasquale Spinelli, Bologna: 24–25, 36 (left), 47, 48, 49, 51, 53, 59, 74, 75 (center), 76–77 (below), 91 (center), 93, 98, 105 (above), 106, 108 (above), 110, 121, 142, 143, 148 (center; below left), 156 (above), 159 (right), 162, 164 (right), 166 (left), 172, 191, 206, 217, 222 (below), 228, 243
Torre Solare di Arcetri (Arcetri Solar Tower): 30 (below), 31 (below)

PICTURE CREDITS

DRAWINGS AND DIAGRAMS:

Centro documentazione Mondadori, Milan: 68 (center)
De Pretto: 68
Studio Pleiadi: 18, 61, 69 (below), 149 (below), 150, 198, 207 (above, center), 213 (right), 221, 228, 231, 235 (below)
Graffito, Milan: 14, 16, 17, 19, 20, 21, 22 (below), 29, 32, 38 (above), 42, 43, 57, 58, 63, 65, 76, 79, 82, 83, 84 (above), 85 (below), 88, 94, 95, 100, 104, 107, 108, 114, 115, 124, 133 (below), 134 (above), 138, 139 (above), 139 (below), 144, 151, 152, 153 (above), 164, 165, 169 (below), 177 (above), 178 (below), 180 (below), 181, 189 (above), 192, 193, 194, 202, 207 (above left), 216 (left), 217 (below), 225, 226, 229 (above), 238

Illustrations in this volume not specifically listed in the credits on this and facing page are from NASA, the Jet Propulsion Laboratory, ESA and ASI.
The editor wishes to apologize for any errors or omissions that may have occurred when acknowledging sources of illustrations.
The editor also wishes to thank the staff of the astronomical review L'Astronomia for all their assistance during the preparation of this book.